Essentials of Stochastic Processes

Translations of
MATHEMATICAL MONOGRAPHS

Volume 231

Essentials of Stochastic Processes

Kiyosi Itô

Translated by Yuji Ito

American Mathematical Society
Providence, Rhode Island

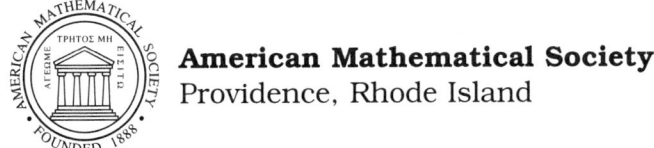

KAKURITSU KATEI, 2 vols. [Essentials of Stochastic Processes]
by Kiyosi Itô © 1957 by Kiyosi Itô

Originally published in Japanese in 1957 by Iwanami Shoten, Publishers, Tokyo.
This English language edition published in 2006 by the American Mathematical Society, Providence, RI, by arrangement with the author c/o Iwanami Shoten, Publishers, Tokyo.

Translated by Yuji Ito

2000 *Mathematics Subject Classification.* Primary 60-02, 60E07, 60G10, 60J25, 60J60, 60G51; Secondary 60G52, 60J35.

For additional information and updates on this book, visit
www.ams.org/bookpages/mmono-231

Library of Congress Cataloging-in-Publication Data

Ito, Kiyosi, 1915-
[Kakuritsu katei. English]
Essentials of stochastic processes / Kiyosi Itô ; translated by Yuji Ito.
p. cm. – (Translations of mathematical monographs ; v. 231)
ISBN 0-8218-3898-9 (alk. paper)
1. Stochastic processes. I. Title. II. Series.

QA274.I86413 2006
519.2′3–dc22 2006042673

∞ The paper used in this book is acid-free and falls within the guidelines
established to ensure permanence and durability.
Visit the AMS home page at http://www.ams.org/

10 9 8 7 6 5 4 3 2 1 11 10 09 08 07 06

Contents

Author's Preface

The present volume, *Essentials of Stochastic Processes*, is an English translation of my book written in Japanese and issued by Iwanami Shoten in 1957 in two parts: Stochastic Processes I (from Chapter 1 to 3) and II (from Chapter 4 to 5). In this work, I provide a unified and comprehensive account of additive processes (or Lévy processes), stationary processes, and Markov processes, which remain to this day the three most important classes of stochastic processes. I had sent the Japanese original at the time of its publication to Eugene B. Dynkin, and I was very pleased to see A. D. Wentzell's Russian translation published in 1960 (Part I) and 1963 (Part II). I am also grateful to Dynkin for editing the translation and adding some important clarification footnotes. In 1959 Shizuo Kakutani at Yale University, noting the significance of my description of the one-dimensional diffusions, advised Yuji Ito, then one of his graduate students, to produce a translation of Part II into English, which was distributed among a limited circle of mathematicians around Yale University as a typewritten mimeograph. On the occasion of my receiving the Kyoto Prize in 1998, Shinzo Watanabe and Masatoshi Fukushima encouraged me to have the entire 1957 book translated into English and published by the American Mathematical Society. Yuji Ito graciously agreed to take on this arduous task and revisited his earlier partial translation, not only adding Part I, but also fully revising his original translation of Part II.

Although almost half a century has passed since the initial publication in Japanese, I hope there is enough of value in this work to merit its publication in English at this time. It should be noted that some detailed introductions to additive processes and Markov processes are given in two of my lecture notes published later on:

- *Lectures on Stochastic Processes*, Tata Institute of Fundamental Research, Bombay, 1960.
- *Stochastic Processes*, edited by Ole E. Barndorff-Nielsen and Ken-iti Sato, Springer, 2004 (originally published as Lecture Notes from Aarhus University in 1969).

However, the present volume is the only one among my English textbooks that includes an introduction to stationary processes.

Chapter 5 is devoted to the one-dimensional diffusion theory which is important as a basic prototype of the study of Markov processes. This chapter starts with a presentation of the local generator of a one-dimensional diffusion process as a generalized second order differential operator discovered by William Feller several years before I wrote this book. It then proceeds to a detailed description of the boundary behaviors of the solutions of the associated homogeneous and inhomogeneous equations in an analytical way, followed by their probabilistic implications on the path properties of the diffusion near the boundaries.

My lecture notes from the Tata Institute mentioned above contain another detailed explanation of the Feller local generator. Section 4.6 of my joint book with H. P. McKean (*Diffusion Processes and Their Sample Paths*, Springer, 1965; in Classics in Mathematics, Springer, 1996) also exhibits the boundary behaviors with some probabilistic proof, while sections 5.12, 5.13, and 5.14 of the present volume are readily understood even by readers unfamiliar with probability theory.

When I wrote the original Japanese version of this book, the real study of stochastic processes had just begun, and not much related literature was available as noted in the Postscript. In the five decades since then, there have been significant developments in the theory of stochastic processes with many important subsequent publications, some of which are listed in the Preface to the Original and the Foreword by the Editors in the above-mentioned *Stochastic Processes* published in 2004 based on my Aarhus Lecture Notes.

I am very much indebted to those who have helped me bring this translation project to a successful completion. My gratitude, first and foremost, goes to Yuji Ito for the precise yet elegant translation which far exceeded my expectations, and I sincerely wish to thank him once again for his time and efforts. My thanks are due to M. Fukushima, K. Ichihara, and S. Watanabe for the meticulous care they took in proof-reading and editing the translated manuscript. This English version is in many ways superior to the original in that it eliminates minor inconsistencies and updates some of the discussion. In particular, the original version in Japanese, written when I had just started my work on paths in Markov processes, contains discussions of the general theory in Chapters 4 and 5 that are in hindsight somewhat unclear and misleading. I am grateful to M. Fukushima and S. Watanabe for suggesting the appropriate amendments in these chapters.

In view of the fact that Professor Shizuo Kakutani had first suggested, shortly after its Japanese publication in 1957, that my book be translated into English, I had hoped to be able to finally present him with this English version published by the American Mathematical Society. It was with great sadness that I learned of his passing away in the summer of 2004 in New Haven. In order to express my deep respect and admiration for his teaching and his important contributions to mathematics, I wish to dedicate this book to the late Professor Shizuo Kakutani.

Kyoto, December 2005 K. Itô

Translator's Foreword

It is my great pleasure to present an English translation of *Essentials of Stochastic Processes* written by Professor Kiyosi Itô. It was almost half a century ago when the original Japanese version of this book was published by Iwanami Shoten. As it is mentioned by Professor Itô in the Author's Preface, I took up the translaton of Part II (Chapters 4 and 5 of the book) into English only a couple of years after the publication of the original with the urging of the late Professor Shizuo Kakutani of Yale University. I was a graduate student in mathematics at Yale at the time, trying to write a Ph.D. thesis under Professor Kakutani's supervision, and he probably thought that I should look into the possibility of working in the field of continuous parameter Markov processes, which was undergoing a rapid development at the time. No doubt, he felt that the best place to follow this development was to read the account by Professor Itô, who was one of the central figures spearheading this development. Professor Kakutani himself was very much interested in the materials contained in this book, and he thought there may be people around Yale and elsewhere in the United States who would benefit a great deal from learning the contents of this book, especially the part on diffusion processes. This was why he urged me to translate (rather than just read through) Part II of the book into English, and when I finished the translation, he decided to have it typed and copies mimeographed by a secretary of the Mathematics Department of Yale and put out as a part of the lecture note series circulated by the Department. I do not know how many copies of the translation were circulated in this manner, but I learned much later that there were a number of people, some of whom eventually became prominent probabilists, who have read the translation and benefitted from it. Although I ended up choosing a thesis topic in Ergodic Theory, a field related but not directly connected with the contents of this book, I certainly learned a great deal about Stochastic Processes in going through the book carefully in the process of translation.

A couple of years ago, Masatoshi Fukushima approached me and asked whether I would be interested in having my old translation (possibly adding a new translation of Part I) published in a more formal manner, as there are materials in it which had never been published in English elsewhere and continue to draw the interest of the specialists in the field. I was delighted to hear this proposal with the additional information that it was the wish of Professor Itô also to have a formal publication of an English translation of this book, and he would like me to take up the task of the actual translation of the entire book. As I was not sure whether Professor Kakutani had asked for permission from Professor Itô to translate the portion of the book before he told me to take up the task and decided to circulate copies of the product through the Mathematics Department of Yale, I was very pleased and honored to learn Professor Itô's wishes, and decided to embark on the new translation project with his blessings.

I had thought that I would be able to finish the project within a year or so, but it took much longer than I had expected, partly because I decided, in addition to translating Part I, to retranslate Part II to make the entire manuscript consistent and easier to read. My lack of any previous experience in writing articles in AMS-LaTeX format also forced me to spend a considerable amount of extra time. I am truly grateful to Fukushima, Shinzo Watanabe, and Kanji Ichihara for proof-reading my manuscript very carefully. Although I tried, while I was translating, to correct minor errors in the original as much as I could, I still missed a few, and furthermore, I introduced new errors, typographical and otherwise, of my own (many of which were caused by my inexperience in AMS-LaTeX typesetting). All of these were found and corrected by Fukushima, Watanabe, and Ichihara. Furthermore, as it was explained by Professor Itô in the Author's Preface, Fukushima and Watanabe suggested a few amendments for arguments used in the original Japanese version, in order to eliminate minor inconsistencies and to update some of the discussion, which would have been impossible for me to do as a non-specialist in the field. I am very happy that with their great help, I was finally able to complete this translation project. I am grateful also to the American Mathematical Society for agreeing to publish this translation of Professor Itô's excellent account of the properties of stochastic processes.

Tokyo, January 2006 Yuji Ito

CHAPTER 1

Basic Concepts

1.1. Measure Theoretic Probability (1) Intuitive Background

Consider a game in which two players A and B take turns to toss a coin and player who comes up with heads first wins the game. Supposing that player A starts off, let us consider the following problems:

(i) What is the probability that A wins the game?
(ii) How many tosses are required on the average to determine the outcome of the game?

Let us consider first what the possibilities are for developments of the game. By designating heads by H and tails by T, we can represent possible developments of the game as

$$(1.1.1)\qquad \left\{\begin{aligned} \omega_1 &= H \\ \omega_2 &= TH \\ \omega_3 &= TTH \\ &\cdots\cdots\cdots \\ \omega_n &= \underbrace{TT\cdots T}_{n-1}H \\ &\cdots\cdots\cdots \\ \omega_\infty &= TTT\cdots \end{aligned}\right\}.$$

Here, ω_1 represents the case where A comes up with heads in the first toss and the game ends, and ω_2 represents the case where A starts with tails and then B comes up with heads ending the game. Likewise, ω_n refers to the case where the first $n-1$ tosses are all tails and on the n-th toss heads appears for the first time ending the game. In the case where n is odd, A wins the game, while if n is even, B is the winner. Finally, ω_∞ refers to the case where both A and B keep on coming up only with tails, and in this case, the game continues forever. We call $\omega_1, \omega_2, \cdots, \omega_\infty$ **sample points** as they represent various samples for possible developments of the game. And the set $\Omega = \{\omega_1, \omega_2, \cdots, \omega_\infty\}$ of all possible sample points will be called the **sample space**.

Let us next figure out the probability of each sample point. Since on the first toss heads and tails can appear with equal likelihood, the probability of ω_1 is $1/2$, and the remaining set of points $\{\omega_2, \omega_3, \cdots, \omega_\infty\}$ should be assigned the probability $1/2$ as well. For the same reason, this probability $1/2$ should be distributed equally between ω_2 and the remaining set $\{\omega_3, \omega_4, \cdots, \omega_\infty\}$. Continuing in this way, we see that the probabilities assigned to points $\omega_1, \omega_2, \omega_3, \cdots, \omega_\infty$ are $1/2, 1/4, 1/8, \cdots, 0$, respectively.

If we denote by $P(\omega)$ the probability assigned to the sample point ω, we see that in the situation under consideration we have

$$P(\omega_1) = 1/2, P(\omega_2) = 1/4, \cdots, P(\omega_n) = 1/2^n, \cdots, P(\omega_\infty) = 0. \tag{1.1.2}$$

Furthermore, if we take a subset E of Ω, the probability assigned to E would be obtained by summing the probabilities assigned to each point of E, namely,

$$P(E) = \sum_{\omega \in E} P(\omega). \tag{1.1.3}$$

In this way we obtain a set function $P(E)$, which we call a **probability distribution**.

Let us next consider problem (ii). The number of tosses required to decide the outcome of the game is determined for each sample point. For example, this number is 1 for ω_1, 2 for ω_2, and n for ω_n. Namely, this is a function defined on the sample space. Suppose we denote this function by $x(\omega)$. A function defined on the sample space such as this one is called a **random variable**. The problem in (ii) mentioned above is to find the average value of a random variable $x(\omega)$. There are several ways of defining the notion of the average of a random variable, but the following one called the **expectation** is used most commonly:

$$E(x) = \sum_{\omega \in \Omega} x(\omega) P(\omega) = \sum_{n=1}^{\infty} n P(\omega_n) = \sum_{n=1}^{\infty} n \cdot \frac{1}{2^n} = 2. \tag{1.1.4}$$

Let us move on to problem (i). The fact that player A wins can be expressed by the statement that $x(\omega)$ mentioned above takes an odd value, namely by the statement "$x(\omega) = \text{odd}$". A fact, such as this one, that can be expressed in terms of some condition concerning sample points ω is called an **event**. The probability of an event is defined to be the probability of the set E consisting of all the sample points satisfying the given condition (such a set is called the **extension** of the event). Consequently, we can state

$$P(\text{A wins}) = P(E) = \sum_{n=0}^{\infty} P(\omega_{2n+1}) = \sum_{n=0}^{\infty} \frac{1}{2^{2n+1}} = \frac{2}{3}. \tag{1.1.5}$$

Let us summarize the essential points of the statements made concerning the example above. Basic concepts are the set Ω called the sample space and the probability distribution P defined on it. A function defined on the set Ω is called a random variable, and its expectation is defined by the equation in (1.1.4). A condition concerning points of Ω is called an event, and the probability of that condition is given by the value $P(E)$ of the distribution P on the extension E of the event. In this connection, there is a problem as to how the probability distribution should be given. If the set Ω is a countable set as in the above-mentioned example, it is sufficient to assign a probability value to each sample point in such a way that the total sum of these values equals 1. However, if Ω is taken to be the set of all points on the real line, or the set of all points in the plane, or more generally, the set of all possible movements of a particle undergoing Brownian motion, there is no naive way, such as above for the case of a countable sample space, to define a probability distribution. However, if you notice that there is a similarity between the notion of a probability distribution and that of a mass distribution, then you may imagine that the measure theory, which gives a rigorous mathematical foundation for mass distributions, may possibly be applied to probability distributions.

This idea happens to be correct, and with the construction of probability theory on the basis of measure theoretic foundations, probability theory has become a truly rigorous mathematical discipline, quite different from what it used to be, namely, collections of ambiguous reasoning and ad hoc methods based on "common sense" and intuition. This new development of probability theory has yielded many valuable and useful applications.

1.2. Probability Distribution

Let X be a set and let $\boldsymbol{B}$ be a Borel field of subsets of X. If a Lebesgue type measure $P(E)$ defined for elements (sets) of $\boldsymbol{B}$ satisfies the condition

$$P(X) = 1, \tag{1.2.1}$$

then P is called a **probability measure**, or a **probability distribution**, or more simply, a **distribution** on $X(\boldsymbol{B})$.

Let us first consider the simplest case where the set X is a finite set. Let us denote the elements of X by $x_1, x_2, \cdots, x_n$. In this case, we usually take $\boldsymbol{B}$ to be the set 2^X of all subsets of X. If the values of probabilities $P(x_1), P(x_2), \cdots, P(x_n)$ are specified, then due to the additivity of the measure, the value $P(E)$ for an arbitrary $E \subseteq X$ is given by

$$P(E) = \sum_{x \in E} P(x). \tag{1.2.2}$$

Consequently, it is enough in this case to specify a point function $P(x)$ to define a distribution. It is clear that the conditions that the function $P(x)$ has to satisfy are

$$P(x) \geq 0, \quad \sum P(x) = 1. \tag{1.2.3}$$

If, in particular, the value of $P(x)$ **does not depend** on x and, consequently, $P(x) = 1/n$ for each x, then P is called the **uniform distribution**.

When the set X is countably infinite, the situation is practically the same as for the case of a finite set, except for the fact that there is no uniform distribution for a countably infinite X.

When X is the set R^1 of real numbers, however, the problem becomes much more difficult. Except for very special cases, it becomes impossible to take $\boldsymbol{B}$ to be 2^{R^1}. The most natural way to choose $\boldsymbol{B}$ is to choose the smallest Borel field $\boldsymbol{B}^1$ containing all open subsets of R^1. Elements of $\boldsymbol{B}^1$ are usually called **Borel sets**. Suppose P is a probability distribution on $R^1(\boldsymbol{B}^1)$. We say that a point $x \in R^1$ belongs to the support of P if for an arbitrary neighborhood U of x, $P(U)$ is positive. The set of all such points is called the **support** of P. If, in particular, $P(x) > 0$, then x is naturally a point of the support of P; we call such a point the **discontinuity point** of P. The set D of all discontinuity points of P is at most countable. If $P(D) = 1$, P is called **purely discontinuous**, while if $P(D) = 0$, then P is said to be **continuous**. There is a notion called **absolute continuity**, which is slightly stronger than the notion of continuity of P. We say that P is absolutely continuous if whenever for a set E the usual Lebesgue measure $|E| = 0$, then $P(E) = 0$. If P is absolutely continuous, then P has the density and can be written as

$$P(E) = \int_E f(x)dx. \tag{1.2.4}$$

The conditions that the density function $f(x)$ must satisfy are

$$f(x) \geq 0, \quad \int_{R^1} f(x)dx = 1. \tag{1.2.5}$$

Probability distribution which is continuous but not absolutely continuous is called **singular**. Purely discontinuous, absolutely continuous, and singular distributions are the three most important types of distributions on $R^1(\boldsymbol{B}^1)$, and furthermore, it is known that an arbitrary distribution can be represented as a **convex combination** of these three types of distributions. (We say that a is a convex combination of $a_1, a_2, \cdots, a_n$ if a can be written as $a = \sum c_i a_i, c_i \geq 0, \sum c_i = 1$.) This fact is known as the **Lebesgue decomposition theorem**.

EXAMPLE 1.2.1. δ(delta)-distribution $\delta(\bullet; a)$. This is a purely discontinuous distribution for which the set D consists of a single point a. The special case where $a = 0$ is called the **unitary distribution**.

EXAMPLE 1.2.2. Binomial distribution $B(\bullet; p, n)$, where $0 < p < 1$, and n is a natural number. This is a purely discontinuous distribution for which $D = \{0, 1, 2, \cdots, n\}$, and

$$P(k) = \binom{n}{k} p^k \cdot q^{n-k}, \quad q = 1 - p, \quad k = 0, 1, 2, \cdots, n. \tag{1.2.6}$$

It is called the binomial distribution since $P(k)$ coincides with the k-th term of the binomial expansion of $(p+q)^n$.

EXAMPLE 1.2.3. Poisson distribution $P(\bullet; \lambda)$, where $\lambda > 0$. This is also a purely discontinuous distribution for which $D = \{0, 1, 2, \cdots\}$, and

$$P(k) = e^{-\lambda} \frac{\lambda^k}{k!}, \quad k = 0, 1, 2, \cdots. \tag{1.2.7}$$

EXAMPLE 1.2.4. Normal distribution $N(\bullet; a, v)$, where a is a real number and $v > 0$. This is an absolutely continuous distribution and its density is given by

$$f(x) = \frac{1}{\sqrt{2\pi v}} e^{-\frac{(x-a)^2}{2v}}. \tag{1.2.8}$$

EXAMPLE 1.2.5. Cauchy distribution $C(\bullet; a, c)$, where a is a real number and $c > 0$. This is also an absolutely continuous distribution and its density is given by

$$f(x) = \frac{c}{\pi} \cdot \frac{1}{c^2 + (x-a)^2}. \tag{1.2.9}$$

When X is the m-dimensional Euclidean space R^m, results for the case of R^1 can be extended almost verbatim. As $\boldsymbol{B}$ we take, just as in the 1-dimensional case, the smallest Borel field $\boldsymbol{B}^m$ containing all open subsets of R^m. Elements of $\boldsymbol{B}^m$ are called Borel sets. The fact that there are three distinct types of distributions and that the Lebesgue decomposition theorem is valid holds true also in R^m.

EXAMPLE 1.2.6. δ-distribution $\delta(\bullet; \boldsymbol{a})$ is defined in the same way as for the 1-dimensional δ-distribution, but with $\boldsymbol{a}$ being an element of R^m.

EXAMPLE 1.2.7. Multinomial distribution $B(\bullet; \boldsymbol{p}, n)$, where n is a natural number, $\boldsymbol{p} = (p_1, p_2, \cdots, p_m)$, $p_i \geq 0$, $\sum p_i = 1$. This is a purely discontinuous distribution for which D is the set of all the m-dimensional lattice points

$\boldsymbol{k} = (k_1, k_2, \cdots, k_m)$, with $\sum k_i = n, \quad k_i \geq 0$, and

$$P(\boldsymbol{k}) = \frac{n!}{k_1!k_2!\cdots k_m!} p_1^{k_1} p_2^{k_2} \cdots p_m^{k_m}. \tag{1.2.10}$$

EXAMPLE 1.2.8. Normal distribution $N(\bullet; \boldsymbol{a}, V)$, where $\boldsymbol{a}$ is an element of R^m and V is a strictly positive definite symmetric matrix. This is an absolutely continuous distribution and its density is given by

$$f(\boldsymbol{x}) = (2\pi)^{-\frac{m}{2}} (\det V)^{-\frac{1}{2}} \exp\{-\frac{1}{2}(V(\boldsymbol{x}-\boldsymbol{a}), (\boldsymbol{x}-\boldsymbol{a}))\}. \tag{1.2.11}$$

Here, $V(\boldsymbol{x}-\boldsymbol{a})$ represents the result of applying the linear transformation V to the vector $\boldsymbol{x}-\boldsymbol{a}$, and $(\quad,\quad)$ denotes the inner product in R^m.

As we outlined above, the extension of results from the case of 1-dimensional space R^1 to the case of m-dimensional space R^m is straightforward, but further extension to the case of infinite-dimensional spaces encounters considerable difficulties. For instance, on infinite-dimensional spaces, there exists no measure corresponding to the usual Lebesgue measure on R^m, and therefore, it is not possible to define the notion of absolute continuity.

When A is an arbitrary set, we consider the set R^A to be the collection of all the real-valued functions ξ defined on the set A arranged as $\prod_\alpha \xi_\alpha$. When the set A is a finite set, then R^A is a finite-dimensional space, but if A is an infinite set, then R^A is infinite dimensional. A mapping defined on R^A which assigns to each point $\prod_\alpha \xi_\alpha$ its "α_0-coordinate" ξ_{α_0} is called a **projection** and is denoted by p_{α_0}. More generally, for n distinct points $\alpha_1, \cdots, \alpha_n$ of A, a mapping on R^A which assigns to each point $\prod_\alpha \xi_\alpha$ the point $(\xi_{\alpha_1}, \xi_{\alpha_2}, \cdots, \xi_{\alpha_n})$ of R^n is also called a projection and is denoted by $p_{\alpha_1\cdots\alpha_n}$. A subset of R^A of the form $p^{-1}_{\alpha_1\cdots\alpha_n}(E^{(n)})$, where $E^{(n)}$ is an n-dimensional Borel set, is called a **Borel cylinder set**. The smallest Borel field of subsets of R^A containing all Borel cylinder sets is denoted by $\boldsymbol{B}^A$, and elements of $\boldsymbol{B}^A$ will be called **Borel subsets** of R^A.

Suppose now P is a distribution on $R^A(\boldsymbol{B}^A)$. For distinct elements $\alpha_1, \alpha_2, \cdots, \alpha_n$ of A, we define a distribution $P_{\alpha_1\cdots\alpha_n}$ on $R^n(\boldsymbol{B}^n)$ in the following way:

$$P_{\alpha_1\cdots\alpha_n}(E^{(n)}) = P(p^{-1}_{\alpha_1\cdots\alpha_n}(E^{(n)})), \quad E^{(n)} \in \boldsymbol{B}^n. \tag{1.2.12}$$

Such distributions $P_{\alpha_1\cdots\alpha_n}$ are called **projections of** P. Let us denote by $\mathfrak{P}$ the collection of all the projections $P_{\alpha_1\cdots\alpha_n}$ of the distribution P. Then, we see that $\mathfrak{P}$ satisfies the following **consistency conditions** of Kolmogorov:

(K.1) if $i(1), i(2), \cdots, i(n)$ is a permutation of $1, 2, \cdots, n$, then
$P_{\alpha_{i(1)}\alpha_{i(2)}\cdots\alpha_{i(n)}}(E_{i(1)} \times E_{i(2)} \times \cdots \times E_{i(n)}) = P_{\alpha_1\cdots\alpha_n}(E_1 \times E_2 \times \cdots \times E_n)$.

(K.2) $P_{\alpha_1\alpha_2\cdots\alpha_n}(E^{(n-1)} \times R^1) = P_{\alpha_1\alpha_2\cdots\alpha_{(n-1)}}(E^{(n-1)})$.

Conversely, if a collection $\mathfrak{P}$ of distributions satisfies the two conditions (K.1) and (K.2) above, then there exists one and only one distribution P on $R^A(\boldsymbol{B}^A)$ which satisfies the condition (1.2.12). This fact is called Kolmogorov's Theorem.

Suppose for each element α of A a probability distribution P_α on $R^1(\boldsymbol{B}^1)$ is given. If we define

$$P_{\alpha_1\cdots\alpha_n} = P_{\alpha_1} \times P_{\alpha_2} \times \cdots \times P_{\alpha_n} \quad \text{(the direct product measure!)}, \tag{1.2.13}$$

then $\mathfrak{P} = \{P_{\alpha_1 \cdots \alpha_n}\}$ satisfies conditions (K.1), (K.2) above, and therefore, according to Kolmogorov's Theorem, this $\mathfrak{P}$ determines a probability distribution P on $R^A(\boldsymbol{B}^A)$. We denote this probability distribution P by $\prod_{\alpha \in A} P_\alpha$ and call it the **direct product probability measure** of $\{P_\alpha, \alpha \in A\}$. It is clear that this P is the distribution on $R^A(\boldsymbol{B}^A)$ characterized by

$$P(p_{\alpha_1}^{-1}(E_1) \cap p_{\alpha_2}^{-1}(E_2) \cap \cdots \cap p_{\alpha_n}^{-1}(E_n)) = \prod_{i=1}^{n} P_{\alpha_i}(E_i). \tag{1.2.14}$$

Similarly, we can define the direct product probability measure for the case where each P_α is a higher-dimensional (can be finite or infinite dimensional, and dimensions may vary according to α) probability distribution.

1.3. Measure Theoretic Probability (2) Mathematical Structure

Fix a set Ω, and call it a sample space. By a **probability space** $\Omega(\boldsymbol{B}, P)$, we mean a triple, consisting of the set Ω, a Borel field $\boldsymbol{B}$ of subsets of Ω, and a probability measure P defined on $\Omega(\boldsymbol{B})$.

As $\Omega(\boldsymbol{B}, P)$ is a measure space of some sort, it is possible to develop Lebesgue integration theory on it. A real-valued measurable function defined on $\Omega(\boldsymbol{B}, P)$ is called a **random variable**. Let $x(\omega)$ be a random variable.

$$\Phi(E) = P(\{\omega / x(\omega) \in E\}) \equiv P(x^{-1}(E)), \quad E \in \boldsymbol{B}^1, \tag{1.3.1}$$

is called the **distribution** of x. It is a probability distribution on $R^1(\boldsymbol{B}^1)$. The quantity

$$E(x) = \int_\Omega x(\omega) P(d\omega) \tag{1.3.2}$$

is called the **expectation** or the **mean** of $x(\omega)$. This quantity can be written in terms of the distribution of x as follows:

$$E(x) = \int_{R^1} \xi \Phi(d\xi). \tag{1.3.3}$$

Since a measurable function may not necessarily be integrable, the mean of a random variable need not exist.

We can define a random vector by listing several random variables together. Let A be either a finite or an infinite set, and suppose to each element α of A there corresponds a random variable x_α. If we define in such a situation

$$\boldsymbol{x} = \prod_{\alpha \in A} x_\alpha, \tag{1.3.4}$$

then $\boldsymbol{x}(\omega)$ becomes a function on Ω taking values in R^A. Furthermore, this function is measurable in the sense that

$$E^{(A)} \in \boldsymbol{B}^A \Rightarrow \boldsymbol{x}^{-1}(E^{(A)}) \in \boldsymbol{B}. \tag{1.3.5}$$

Such an $\boldsymbol{x}(\omega)$ will be called a random vector. When the cardinality of the set A is m, it is called an m-dimensional random vector. A 2-dimensional random vector $(x_1(\omega), x_2(\omega))$ is sometimes represented as $x_1(\omega) + i x_2(\omega)$ and called a **complex random variable**. The distribution of a random vector is also defined by formula (1.3.1), and the **mean vector** for a random vector can be defined by taking the integrals componentwise.

A mapping φ from R^A to R^B which satisfies

$$E^{(B)} \in \boldsymbol{B}^B \Rightarrow \varphi^{-1}(E^{(B)}) \in \boldsymbol{B}^A \tag{1.3.6}$$

is said to be **Borel measurable** or ***B*-measurable** or more simply **measurable**. The image of a random vector under a measurable map is measurable. Namely, if we apply a measurable map φ to a random vector $\boldsymbol{x}(\omega)$ taking values in R^A, then we get a random vector $\varphi(\boldsymbol{x}(\omega))$ taking values in R^B. The mean vector of the random vector $\varphi(\boldsymbol{x}(\omega))$ is given by

$$E[\varphi(\boldsymbol{x}(\omega))] = \int_{R^A} \varphi(\xi)\Phi(d\xi), \quad \Phi \text{ is the distribution of } \boldsymbol{x}(\omega). \tag{1.3.7}$$

$\varphi(\boldsymbol{x}(\omega))$ is said to be **measurable with respect to** $\boldsymbol{x}(\omega)$.

When there is a family $\boldsymbol{x}_\alpha(\omega)$, $\alpha \in A$, of random vectors, we can obtain a higher-dimensional random vector by considering

$$\boldsymbol{x}(\omega) = \prod_\alpha \boldsymbol{x}_\alpha(\omega). \tag{1.3.8}$$

If the distribution of $\boldsymbol{x}(\omega)$ is the direct product of the distributions of $\boldsymbol{x}_\alpha(\omega)$, $\alpha \in A$, then $\boldsymbol{x}_\alpha, \alpha \in A$, is said to be an **independent** family. This condition of independence is characterized by the following property:

Whenever $\alpha_1, \alpha_2, \cdots, \alpha_n$ are distinct elements of A, then

$$P\{\omega/\boldsymbol{x}_{\alpha_i} \in E_i, \ i = 1, 2, \cdots, n\} = \prod_{i=1}^n P\{\omega/\boldsymbol{x}_{\alpha_i}(\omega) \in E_i\}. \tag{1.3.9}$$

When $A = \bigcup_\lambda A_\lambda$ (disjoint union), and $\boldsymbol{x}_\alpha$, $\alpha \in A$, is independent, then

$$\boldsymbol{y}_\lambda(\omega) = \prod_{\alpha \in A_\lambda} \boldsymbol{x}_\alpha(\omega), \qquad \lambda \in \Lambda, \tag{1.3.10}$$

is independent as well. Also, if $\boldsymbol{x}_\alpha(\omega)$, $\alpha \in A$, is independent, and if for each α, φ_α is a measurable map, then $\varphi_\alpha(\boldsymbol{x}_\alpha(\omega))$, $\alpha \in A$, is independent.

When $x_1(\omega), x_2(\omega), \cdots, x_n(\omega)$ is an independent sequence of complex (or real) random variables, then

$$E[x_1(\omega) \cdots x_n(\omega)] = E[x_1(\omega)]E[x_2(\omega)] \cdots E[x_n(\omega)] \tag{1.3.11}$$

holds. This is called the **multiplicativity** of the mean.

There are several ways of defining the notion of convergence of a sequence of random variables (or random vectors) $x_n(\omega)$, $n = 1, 2, \cdots$, to $x(\omega)$. One of these is a most natural one given by

$$P\{\omega/|x_n(\omega) - x(\omega)| \to 0\} = 1. \tag{1.3.12}$$

Here, $| \ |$ denotes the length of a vector. $|x_n(\omega) - x(\omega)|$ is a measurable function of ω and since

$$\{\omega/|x_n(\omega) - x(\omega)| \to 0\} = \bigcap_p \bigcup_N \bigcap_{n>N} \{\omega/|x_n(\omega) - x(\omega)| < \frac{1}{p}\}, \tag{1.3.13}$$

the set $\{\omega/\cdots\}$ appearing on the left-hand side of (1.3.12) is a measurable set. Equation (1.3.12) states that the P-measure of this measurable set equals 1. The convergence defined in this way is called **almost everywhere convergence** on Ω, or **convergence with probability** 1, or **almost sure convergence**, and denoted by $x_n \to x$ (a.e.) *.

There is another notion of convergence weaker than the one mentioned above. Namely, we stipulate that for every $\epsilon > 0$

$$P\{\omega/|x_n(\omega) - x(\omega)| > \epsilon\} \to 0. \tag{1.3.14}$$

This mode of convergence is called the **convergence in probability** and is denoted by $x_n(\omega) \to x$ (P). If this takes place, then we can show that the distribution of x_n converges in the sense we shall discuss in the next section to the distribution of x. We also consider the notion of convergence given by

$$E(|x_n - x|^p) \to 0 \tag{1.3.15}$$

if $E(|x_n - x|^p)$ is finite for each n. This mode of convergence is called the **mean convergence of the p-th power**. The case where $p = 2$ is considered most frequently, and is called simply the **mean convergence**. This mode of convergence is stronger than the convergence in probability.

1.4. Distribution Function, Characteristic Function, Mean, Variance

Let us denote by $\Phi, \Phi_1, \Phi_2, \cdots$ the probability distributions on $R^1(\mathbf{B}^1)$. The function defined on R^1 by

$$F(\xi) = \Phi(-\infty, \xi] \tag{1.4.1}$$

is called the **distribution function** of Φ. Conditions to be satisfied by a distribution function $F(\xi)$ are the following:

(i) non-decreasing property: $\xi < \eta \Rightarrow F(\xi) \leq F(\eta)$,

(ii) right continuity: $F(\xi + 0) = F(\xi)$,

(iii) $F(-\infty) = 0, \quad F(\infty) = 1$.

Conversely, if a function $F(\xi)$ satisfies these conditions, then it is the distribution function of a distribution Φ on $R^1(\mathbf{B}^1)$ given by

$$\Phi(E) = \int_E dF(\xi) \quad \text{(Lebesgue-Stieltjes Integral)}. \tag{1.4.2}$$

Let $x(\omega)$ be a random variable. The distribution function $F(\xi)$ of the distribution Φ of $x(\omega)$ is given by

$$F(\xi) = P\{\omega/x(\omega) \leq \xi\}. \tag{1.4.3}$$

This function is also called the **distribution function of** $x(\omega)$.

Since probability distributions Φ and distribution functions F correspond in one-to-one fashion as we explained above, we can represent a probability distribution by the corresponding distribution function. Let us denote distribution functions by $F, F_1, F_2, \cdots$ corresponding to probability distributions $\Phi, \Phi_1, \Phi_2, \cdots$, respectively. We define the notion of convergence of a sequence $\{\Phi_n\}$ of probability distributions in the following way: we say that $\Phi_n \to \Phi$ if

$$F_n(\xi) \to F(\xi) \text{ holds at every continuity point } \xi \text{ of } F(\xi). \tag{1.4.4}$$

*a.e. is an abbreviation of "almost everywhere".

This condition is known to be equivalent to either one of the following two conditions: At every point ξ_m belonging to some dense subset $\{\xi_m\}$ of R^1 the following inequalities hold:

$$F(\xi_m - 0) \leq \underline{\lim}_n F_n(\xi_m) \leq \overline{\lim}_n F_n(\xi_m) \leq F(\xi_m + 0) \tag{1.4.4$'$}$$

or for an arbitrary bounded continuous function $f(\xi)$ on R^1,

$$\int_{R^1} f(\xi)\Phi_n(d\xi) \to \int_{R^1} f(\xi)\Phi(d\xi). \tag{1.4.4$''$}$$

It can be shown also that if

$$\inf_n \Phi_n[-a, a] \to 1 \qquad (a \to \infty), \tag{1.4.5}$$

then we can choose a suitable subsequence of Φ_n which will converge to some distribution.

The Fourier transform φ of Φ, namely

$$\varphi(z) = \int_{R^1} e^{iz\xi}\Phi(d\xi), \qquad -\infty < z < \infty, \tag{1.4.6}$$

is called the **characteristic function** of Φ. If Φ is the distribution of a real-valued random variable $x(\omega)$, then

$$\varphi(z) = E(e^{izx}) \tag{1.4.7}$$

and this is called the characteristic function of $x(\omega)$. It is clear that the characteristic function $\varphi(z)$ satisfies the following conditions:

$$\varphi(0) = 1, \quad |\varphi(z)| \leq 1, \tag{1.4.8}$$

$$\varphi(z) \text{ is uniformly continuous on } -\infty < z < \infty. \tag{1.4.9}$$

Positive definiteness: For an arbitrary choice of complex numbers $a_1, a_2, \cdots, a_n$ and of real numbers $z_1, z_2, \cdots, z_n$

$$\sum_{i,j} a_i \bar{a}_j \varphi(z_i - z_j) \geq 0. \tag{1.4.10}$$

Conversely, a positive definite function $\varphi(z)$ satisfying the condition $\varphi(0) = 1$ and continuous at $z = 0$ is a characteristic function of some probability distribution Φ. This fact is known as **Bochner's Theorem**.

Let us denote by $\varphi, \varphi_1, \varphi_2, \cdots$ characteristic functions of $\Phi, \Phi_1, \Phi_2, \cdots$, respectively. Relations between Φ and φ exhibited below are obtained mainly by P. Lévy.

(i) $\varphi_1(z) \equiv \varphi_2(z) \Longleftrightarrow \Phi_1 = \Phi_2$.
(ii) $\varphi_n(z) \to \varphi(z)$ for all $z \Longleftrightarrow \Phi_n \to \Phi$.
(iii) If $\varphi_n(z)$ converges to some function $\theta(z)$ (it is not necessary to assume that $\theta(z)$ is a characteristic function of some distribution) at every z, and the convergence is uniform in some neighborhood of $z = 0$, then $\theta(z)$ is a characteristic function of some distribution Φ. Consequently, it follows from (ii) that $\Phi_n \to \Phi$ holds as well.

The **mean** and the **variance** of a distribution Φ are defined respectively by

$$M(\Phi) = \int_{R^1} \xi\Phi(d\xi), \quad V(\Phi) = \int_{R^1} (\xi - M(\Phi))^2\Phi(d\xi). \tag{1.4.11}$$

The mean is regarded as the central value of the distribution, and the variance can be regarded as a quantity which indicates how the distribution is scattered. If Φ is the distribution of a random variable $x(\omega)$, then $M(\Phi)$ equals $E(x)$, and $V(\Phi)$ equals $E[(x - E(x))^2]$. The latter quantity is also written $V(x)$, and is called the variance of x. It is clear that

$$V(x) = E(x^2) - E(x)^2. \tag{1.4.12}$$

Suppose x_1, x_2 are independent random variables and let x be their sum. Let us denote by Φ_1, Φ_2, Φ the distributions of x_1, x_2, x, respectively, and let F_1, F_2, F be the corresponding distribution functions, $\varphi_1, \varphi_2, \varphi$ the corresponding characteristic functions, and V_1, V_2, V the corresponding variances, respectively. Then we have

$$\Phi(M) = \int_{R^1} \Phi_1(M - \xi)\Phi_2(d\xi) = \int_{R^1} \Phi_2(M - \xi)\Phi_1(d\xi), \tag{1.4.13}$$

where $M - \xi = \{\eta - \xi / \eta \in M\}$,

$$F(\xi) = \int_{R^1} F_1(\xi - \eta)dF_2(\eta) = \int_{R^1} F_2(\xi - \eta)dF_1(\eta), \tag{1.4.14}$$

$$\varphi(z) = \varphi_1(z)\varphi_2(z) \quad \text{(multiplicativity of characteristic function)}, \tag{1.4.15}$$

$$V = V_1 + V_2 \quad \text{(additivity of variance)}. \tag{1.4.16}$$

To prove the first identity of (1.4.13), let c_M be the indicator function of the set M. Then since the distribution of the random vector (x_1, x_2) is given by $\Phi_1 \times \Phi_2$, we get

$$\begin{aligned}\Phi(M) &= E[c_M(x)] = E[c_M(x_1 + x_2)] \\ &= \int_{R^1}\int_{R^1} c_M(\xi_1 + \xi_2)\Phi_1(d\xi_1)\Phi_2(d\xi_2) \\ &= \int_{R^1} \Phi_1(M - \xi_2)\Phi_2(d\xi_2).\end{aligned}$$

Other identities can be established in a similar manner.

Apart from random variables, suppose we define the set function Φ by (1.4.13) for a pair of arbitrary distributions Φ_1, Φ_2; then Φ gives a distribution as well. We denote this distribution by $\Phi_1 * \Phi_2$ and call it the convolution of Φ_1 and Φ_2. If we let $\varphi_1, \varphi_2, \varphi$ be the characteristic functions of Φ_1, Φ_2, Φ, respectively, then we have

$$\varphi(z) = \varphi_1(z)\varphi_2(z) \Longleftrightarrow \Phi = \Phi_1 * \Phi_2. \tag{1.4.17}$$

Let us define the **reflection** $\check{\Phi}$ of a distribution Φ by

$$\check{\Phi}(M) = \Phi(-M), \quad -M = \{-\xi / \xi \in M\}. \tag{1.4.18}$$

The characteristic function $\check{\varphi}$ of $\check{\Phi}$ is given by

$$\check{\varphi}(z) = \overline{\varphi(z)}. \tag{1.4.19}$$

If the distribution of a random variable x is Φ, then the distribution of $-x$ is $\check{\Phi}$. If the distribution Φ is a convex combination of the distributions $\{\Phi_i\}$, then φ is a convex combination of $\{\varphi_i\}$ with the same coefficients. From these facts we can see that the set $\mathbf{C}$ of all characteristic functions satisfies the following properties:

(i) The set $\mathbf{C}$ is invariant under the formation of convex combinations,
(ii) $\mathbf{C} \ni \varphi_1, \varphi_2 \Rightarrow \mathbf{C} \ni \varphi_1 \cdot \varphi_2$
(iii) $\mathbf{C} \ni \varphi \Rightarrow \mathbf{C} \ni \overline{\varphi}$, and therefore,

(iv) $\mathbf{C} \ni \varphi \Rightarrow \mathbf{C} \ni |\varphi|^2$.

If $\varphi(z)$ is the characteristic function of a random variable x, then the characteristic function of $ax + b$ is given by $e^{ibz}\varphi(az)$.

Let us compute the mean M, the variance V, and the characteristic function $\varphi(z)$ for 1-dimensional distributions mentioned as examples in §1.2.

TABLE 1

Φ		M	V	$\varphi(z)$
δ-distribution	$\delta(\bullet; a)$	a	0	e^{iaz}
binomial distribution	$B(\bullet; p, n)$	np	npq	$(pe^{iz} + q)^n$
Poisson distribution	$P(\bullet; \lambda)$	λ	λ	$\exp\{\lambda(e^{it} - 1)\}$
normal distribution	$N(\bullet; a, v)$	a	v	$\exp\left\{iaz - \frac{v}{2}z^2\right\}$
Cauchy distribution	$C(\bullet; a, c)$	undefined	undefined	$\exp\{iaz - c\|z\|\}$

Since $(pe^{iz} + q)^n \cdot (pe^{iz} + q)^{n'} = (pe^{iz} + q)^{n+n'}$, it follows from (1.4.17) that
$$B(\bullet; p, n) * B(\bullet; p, n') = B(\bullet; p, n + n').$$
Similarly, we have
$$\delta(\bullet; a) * \delta(\bullet; a') = \delta(\bullet; a + a'),$$
$$P(\bullet; \lambda) * P(\bullet; \lambda') = P(\bullet; \lambda + \lambda'),$$
$$N(\bullet; a, v) * N(\bullet; a', v') = N(\bullet; a + a', v + v'),$$
$$C(\bullet; a, c) * C(\bullet; a', c') = C(\bullet; a + a', c + c').$$

The results described above for 1-dimensional distributions can be extended verbatim to the case of m-dimensional distributions. The m-dimensional distribution function is defined by

$$F(\xi_1, \xi_2, \cdots, \xi_m) = \Phi((-\infty, \xi_1] \times (-\infty, \xi_2] \times \cdots \times (-\infty, \xi_m]). \tag{1.4.20}$$

The definition of the convergence of the sequence of distributions is also the same as for the 1-dimensional case. The characteristic function is defined by

$$\varphi(z) = \int_{R^m} e^{i(z,\xi)}\Phi(d\xi), \quad (z, \xi) = \sum_{\nu=1}^{m} z_\nu \xi_\nu, \tag{1.4.21}$$

and its properties are exactly the same as for the 1-dimensional case. The mean and the variance become the mean vector and the covariance matrix, respectively. Their components are given by

$$M_i = \int_{R^m} \xi_i \Phi(d\xi), \quad V_{i,j} = \int_{R^m} (\xi_i - M_i)(\xi_j - M_j)\Phi(d\xi). \tag{1.4.22}$$

If Φ is the distribution of a random vector $\boldsymbol{x} = (x_1, x_2, \cdots, x_m)$, then the quantities F, φ, M, V mentioned above become
$$F(\xi_1, \cdots, \xi_m) = P\{\omega / x_1(\omega) \leq \xi_1, \cdots, x_m(\omega) \leq \xi_m\},$$
$$\varphi(z) = E(e^{i(z,\xi)}),$$
$$M = E(\boldsymbol{x}), \quad V_{i,j} = E\{(x_i - E(x_i))(x_j - E(x_j))\}.$$

M, V, φ for the examples of m-dimensional distributions mentioned in §1.2 are tabulated in Table 2.

TABLE 2

Φ	M	V	φ
$\delta(\bullet;\boldsymbol{a})$	$\boldsymbol{a}$	0	$e^{i(z,\boldsymbol{a})}$
$B(\bullet;\boldsymbol{p},n)$	$n\cdot\boldsymbol{p}$	$\begin{cases} V_{\mu\mu}=np_\mu(1-p_\mu) \\ V_{\mu\nu}=-np_\mu p_\nu \quad (\mu\neq\nu)\end{cases}$	$(\sum_\nu p_\nu e^{iz_\nu})^n$
$N(\bullet;\boldsymbol{a},V)$	$\boldsymbol{a}$	V	$\exp\{i(\boldsymbol{a},z)-\frac{1}{2}(Vz,z)\}$

When we defined the m-dimensional normal distribution, we assumed that V is a strictly positive definite, symmetric matrix. If V is positive definite in the wide sense and symmetric, then if we let $V_n = V + I/n$, where I is the identity matrix, V_n becomes a strictly positive definite, symmetric matrix, and therefore the normal distribution $N_n = N(\bullet;\boldsymbol{a},V_n)$ is well defined. The characteristic function $\varphi_n(z)$ of the distribution N_n is given by

$$\varphi_n(z) = \exp\{i(\boldsymbol{a},z) - \frac{1}{2}(V_n z, z)\} = \varphi(z)\exp\{-\frac{1}{2n}(z,z)\},$$

where

$$\varphi(z) = \exp\{i(\boldsymbol{a},z) - \frac{1}{2}(Vz,z)\}.$$

It is easy to see that as $n\to\infty$, $\varphi_n(z)$ converges to $\varphi(z)$ uniformly on any bounded set of z. Since $\varphi_n(z)$ is the characteristic function of N_n, $\varphi(z)$ is also a characteristic function of some distribution N, and furthermore, N is the limit of the sequence $\{N_n\}$ of distributions. We denote this N by $N(\bullet;\boldsymbol{a},V)$ and call it also a normal distribution. This definition of the normal distribution coincides with the definition previously given for the case where V is strictly positive definite and symmetric. If V is not strictly positive definite, namely, if $\det V = 0$, then we call this distribution a **degenerate normal distribution**. In such a case the support of the distribution is not the whole space R^m, but is some hyperplane contained in R^m. Even when the distribution is degenerate, $\boldsymbol{a}$ and V are the mean and the covariance matrix of $N(\bullet;\boldsymbol{a},V)$, respectively.

Finally, let us define the **mean vector** M, the **covariance matrix** V, and the **characteristic function** $\varphi(z)$ for a distribution on an infinite-dimensional space $R^A(\boldsymbol{B}^A)$. The definitions of M and V are exactly the same as for the m-dimensional case. However, there is a slight difference in the case of the characteristic function. While the characteristic function in the m-dimensional case was defined for an arbitrary vector z in R^m, in the case of an infinite-dimensional R^A, we take as the domain of definition of a characteristic function the set of all points z of R^A which equal zero at almost all coordinates, i.e., $z_\alpha = 0$ for all but a finite number of α's in A. Let us denote by R_0^A the set of all such points of R^A. For any pair $z\in R_0^A$ and $\xi\in R^A$, we can define the inner product $(z,\xi)=\sum_\alpha z_\alpha\xi_\alpha$. Since the right-hand side of this identity is actually a finite sum, there is no problem of convergence. We define the characteristic function $\varphi(z)$ of a distribution P on $R^A(\boldsymbol{B}^A)$ by

$$\varphi(z) = \int_{R^A} e^{i(z,\xi)}P(d\xi), \quad z\in R_0^A.$$

Fix a finite number of distinct elements of A. For $\varphi(z)$, we get a function by letting z_α to be zero for all α except for the fixed ones. This function will be called the **section** of $\varphi(z)$ determined by the fixed finite subset of A. A section of the characteristic function of a distribution P is the characteristic function of the projection of P defined in (1.2.12). **Kolmogorov's Theorem** mentioned in §1.2 can be reformulated as follows:

If an arbitrary section of a function $\varphi(z)$ defined on R_0^A is a characteristic function, then $\varphi(z)$ is the characteristic function of a distribution on $R_A(\boldsymbol{B}^A)$. This distribution is uniquely determined.

We can define the **infinite-dimensional normal distribution** in the following way by using this reformulated version of Kolmogorov's Theorem.

Let M be an arbitrary element of R^A, and let V be an element of $R^{A\times A}$ satisfying the following properties:

$$V_{\alpha\beta} = V_{\beta\alpha} ,$$
$$(Vz, z) = \sum_{\alpha\beta} V_{\alpha\beta} z_\alpha z_\beta \geq 0 \quad (z \in R_0^A).$$

(Note that the right-hand side of the second equation above is also a finite sum.)

If we define

$$\varphi(z) = \exp\{i(z, M) - \tfrac{1}{2}(Vz, z)\},$$

then an arbitrary section of $\varphi(z)$ is a distribution function of a normal distribution (possibly degenerate), and therefore, by the reformulated Kolmogorov Theorem, $\varphi(z)$ determines a distribution on $R^A(\boldsymbol{B}^A)$. This distribution is called the normal distribution $N(\bullet; M, V)$ on R^A. The projection of this distribution onto $(\alpha_1, \alpha_2, \cdots, \alpha_n)$ is the distribution corresponding to the section of $\varphi(z)$ on $(\alpha_1, \alpha_2, \cdots, \alpha_n)$, and therefore, it coincides with $N(\bullet; (M_{\alpha_i}), (V_{\alpha_i\alpha_j}))$. From these facts, it follows easily that M and V are the mean and the covariance matrix of this distribution $N(\bullet; M, V)$, respectively.

1.5. Stochastic Process

Stochastic process or **random process** is a notion describing abstractly quantities occurring by chance and changing with the passage of time. From the stand point of the measure theoretic probability theory, this notion is formulated in the following way. Let $\Omega(\boldsymbol{B}, P)$ be the basic probability space and let T be a set of real numbers. A family of random variables $x_t(\omega)$, parametrized by $t \in T$, is called a stochastic process. For applications, t refers to time, and $x_t(\omega)$ represents the value of a quantity occurring by chance at time t. T may be taken to be either a discrete set such as $\{1, 2, 3, \cdots\}$ or $\{\cdots, -3, -2, -1, 0, 1, 2, \cdots\}$, or a continuous set such as intervals $(0, \infty), (-\infty, \infty)$, or (a, b). In case T is discrete, the stochastic process is usually called a **random sequence**.

A stochastic process $x_t(\omega), t \in T$, can be regarded as an infinite-dimensional random vector $\boldsymbol{x}(\omega) = \prod_t x_t(\omega)$. $\boldsymbol{x}(\omega)$ is a random vector taking values in $R^T(\boldsymbol{B}^T)$, and since the elements of the set R^T are real-valued functions defined on T, each ω determines a real-valued function on T. In this sense a stochastic process is sometimes called a **random function**. The function of t determined by ω is called a **sample function** or a sample process. The distribution of a stochastic process regarded as a random vector taking values in $R^T(\boldsymbol{B}^T)$ is a probability distribution on $R^T(\boldsymbol{B}^T)$. When this distribution is a normal distribution, the stochastic process is called a **normal stochastic process**. When $x_t(\omega)$ is a complex-valued measurable function of ω, $x_t(\omega), t \in T$, is called a complex stochastic process. We will discuss later in §3.6 a notion generalizing a normal stochastic process to the case of complex stochastic processes.

CHAPTER 2

Additive Processes

2.1. Definition of Additive Process

An **additive process** or a **differential process** is a stochastic process which arises when independent increments are added as time goes on. A more precise definition is given as follows. A stochastic process $x_t(\omega), a \le t < b\ (-\infty < a < b \le \infty)$, is called an additive process if the following two conditions are satisfied:

(i) $x_a(\omega) \equiv 0$.

(ii) For any choice of $a \le t_0 < t_1 < \cdots < t_n < b,\ x_{t_i} - x_{t_{i-1}}, i = 1, 2, \cdots, n$, are independent.

We define a random sequence $x_n(\omega), n = 0, 1, 2, \cdots$, to be an **additive sequence** in a similar manner. In this case, condition (ii) above is replaced by the requirement: $y_n = x_n - x_{n-1}, n = 1, 2, \cdots$, is an independent sequence of random variables. The following **Fundamental Construction Theorem** will be the basis for subsequent discussions of additive processes.

THEOREM 2.1.1. *For an arbitrary sequence of* 1*-dimensional distributions* Φ_1, $\Phi_2, \cdots$, *one can construct an additive sequence* $\{x_n\}$ *on a suitable probability space in such a way that the distribution of* $x_n - x_{n-1}$ *is given by* Φ_n *for each* n.

PROOF. If the desired sequence $\{x_n\}$ can be constructed, then $\{y_n = x_n - x_{n-1}\}, n = 1, 2, \cdots$, becomes an independent sequence of random variables and for each n the distribution of y_n is given by Φ_n. Consequently, the distribution of the infinite-dimensional random vector $\boldsymbol{y}(\omega) = \prod_i y_i(\omega)$ is the direct product of $\Phi_n, n = 1, 2, \cdots$. Noting this fact, we can proceed as follows. Let $T = \{1, 2, \cdots\}$, $\Omega = R^T, \boldsymbol{B} = \boldsymbol{B}^T, P = \prod_n \Phi_n$, and for $\omega \in \Omega$, let

$$y_n(\omega) = p_n(\omega), \quad x_n(\omega) = \sum_{\nu=1}^{n} y_\nu(\omega). \tag{2.1.1}$$

$\{x_n\}$ obtained in this way is the desired additive sequence. □

Let us derive a corresponding theorem for the case of an additive process. Let $x_t, a \le t < b$, be an additive process, and let Φ_{st} be the distribution of $x_t - x_s$ $(s < t)$. If $s < t < u$, then $x_u - x_s$ is the sum of two independent random variables $x_t - x_s$ and $x_u - x_t$, and therefore,

$$\Phi_{su} = \Phi_{st} * \Phi_{tu} \quad (s < t < u). \tag{2.1.2}$$

THEOREM 2.1.2. *If a family* Φ_{st}, $a \le s < t < b$, *of distributions satisfies the condition* (2.1.2), *then an additive process* x_t, $a \le t < b$, *can be constructed on a suitable probability space in such a way that the distribution of* $x_t - x_s$ *is given by* Φ_{st}.

PROOF. In order to obtain a clue for a possible method of construction of such a process, let us suppose that the desired x_t was obtained, and try to calculate the characteristic function φ of the infinite-dimensional random vector $\boldsymbol{x}(\omega) = \prod_t x_t(\omega)$. Let $T = [a, b)$. For $z \in R_0^T$, let

$$\varphi(z) = E(e^{i(z, \boldsymbol{x})}). \tag{2.1.3}$$

If the coordinates of z are all 0 except for $z_{t_1}, z_{t_2}, \cdots, z_{t_n}$ $(t_1 < t_2 < \cdots < t_n)$, then

$$\begin{aligned}\varphi(z) &= E(exp\{i \sum_\nu z_{t_\nu} x_{t_\nu}\}) \\ &= E(exp\{i \sum_\nu z'_\nu (x_{t_\nu} - x_{t_{\nu-1}})\}) \quad (z'_\nu = \sum_{i=\nu}^{n} z_{t_i}, \ t_0 = 0) \\ &= \prod_\nu \varphi_{t_{\nu-1}t_\nu}(z'_\nu) \quad (\varphi_{st} \text{ is the characteristic function of } \Phi_{st})\end{aligned}$$

due to the independence of $\{x_{t_\nu} - x_{t_{\nu-1}}\}$. Consequently,

$$\varphi(z) = \prod_\nu \varphi_{t_{\nu-1}t_\nu}(z_{t_\nu} + z_{t_{\nu+1}} + \cdots + z_{t_n}). \tag{2.1.4}$$

Thus it has become clear how the distribution of $\boldsymbol{x}(\omega)$ should be defined. So, let us proceed with the proof of the theorem. For $z \in R_0^T$ let us define $\varphi(z)$ by means of (2.1.4). In connection with this definition, we have to note the following fact:

If, for example, z_{t_ν} among $z_{t_1}, z_{t_2}, \cdots, z_{t_n}$ happens to be 0, then $\varphi(z)$ can also be defined using only $\varphi_{t_0t_1}, \varphi_{t_1t_2}, \cdots, \varphi_{t_{\nu-2}t_{\nu-1}}, \varphi_{t_{\nu-1}t_{\nu+1}}, \varphi_{t_{\nu+1}t_{\nu+2}}, \cdots, \varphi_{t_{n-1}t_n}$, but then we have to make sure that the value of $\varphi(z)$ defined in that way coincides with the quantity on the right-hand side of (2.1.4) with $z_{t_\nu} = 0$ in order that our definition makes sense.

In order to assure ourselves that this point causes no problem, it suffices to show that $\varphi_{t_{\nu-1}t_{\nu+1}}(z) = \varphi_{t_{\nu-1}t_\nu}(z)\varphi_{t_\nu t_{\nu+1}}(z)$, but this fact readily follows from the hypothesis (2.1.2).

Next, we have to show that the right-hand side of (2.1.4) as a function of $z_{t_1}, z_{t_2}, \cdots, z_{t_n}$ is a characteristic function of some n-dimensional distribution. For this purpose, we let a probability space $\Omega'(\boldsymbol{B}', P')$ be the space $R^n(\boldsymbol{B}^n)$ together with the direct product probability distribution $\prod_\nu \Phi_{t_{\nu-1}t_\nu}$ defined on it, and consider random variables $y'_\nu(\omega') = p_\nu(\omega')$, $\nu = 1, 2, \cdots, n$. Then, their distributions are given by $\Phi_{t_{\nu-1}t_\nu}$, respectively, and they are independent. Consequently, if we let $x'_\nu(\omega') = \sum_{i=1}^{\nu} y'_i(\omega')$, and define $\varphi'(z_1, z_2, \cdots, z_n)$ to be the characteristic function of the distribution of the random vector $\boldsymbol{x}'(\omega') = \prod_\nu x'_\nu(\omega')$, then we see that the right-hand side of (2.1.4) coincides with $\varphi'(z_{t_1}, z_{t_2}, \cdots, z_{t_n})$. Therefore, by using the reformulated version of the Kolmogorov Theorem, we can conclude that $\varphi(z)$ is a characteristic function of some distribution P on $R^T(\boldsymbol{B}^T)$. By letting $\Omega = R^T, \boldsymbol{B} = \boldsymbol{B}^T$, and defining $x_t(\omega) = p_t(\omega)$ for $\omega \in \Omega$, we get the desired additive process $\{x_t\}$ on the probability space $\Omega(\boldsymbol{B}, P)$. □

2.2. Examples of Additive Processes

Suppose we keep on tossing coins indefinitely, and let x_n be the number of times heads comes up by the n-th toss ($x_0 = 0$), and let $y_1 = x_1, y_2 = x_2 - x_1, y_3 = x_3 - x_2, \cdots$. Then y_n takes the value 1 or 0 depending on whether the n-th toss is heads or tails. Consequently, for each n, the distribution Φ_n of y_n is a purely discontinuous distribution giving probability $1/2$ each to 0 and 1. Furthermore, $\{y_n\}$ is a sequence of independent random variables, and therefore, $\{x_n\}$ is an additive sequence, and its mathematical model can be constructed by using Theorem 2.1.1.

Next, let us consider the problem of **random walk**. Let us suppose a drunken man starts at some point on a road stretching east and west, and takes each step going east or west completely by chance. Let us denote by x_n the position he is on after n steps (positive if the position lies to the east of the starting point, and negative if it lies to the west), and let $y_1 = x_1, y_2 = x_2 - x_1, \cdots$. Then y_n takes the value ± 1 depending on whether he goes east or west at the n-th step, and each possibility has probability $1/2$. Namely, the distribution Φ_n of y_n is the purely discontinuous distribution $\Phi_n(\pm 1) = 1/2$. Also, we may consider $\{y_n\}$ to be independent, and therefore, $\{x_n\}$ is an additive sequence, and its mathematical model can be constructed also by using Theorem 2.1.1.

Next, let us consider **Poisson process** as an example of additive process. This is an additive process x_t, $0 \leq t < \infty$, for which the distribution Φ_{st} of $x_t - x_s$ $(t > s)$ is given by the Poisson distribution $P(\bullet; \lambda(t-s))$, λ being some positive constant. Φ_{st} satisfies condition (2.1.2) of Theorem 2.1.2 as we saw in §1.4, and therefore, we can apply this theorem. From the definition of Poisson distribution, it easily follows that

$$\text{as } h \to 0, \quad P\{\omega/x_{t+h} - x_t \geq 1\} \sim P\{\omega/x_{t+h} - x_t = 1\} \sim \lambda \cdot h \quad \text{holds.}$$

Suppose certain phenomena keep on occurring haphazardly as time goes on, and the occurrences of phenomena over disjoint time intervals are independent, and furthermore, the probability of occurrence in the time interval $(t, t + dt)$ equals $\lambda \cdot dt$. Then the number x_t of occurrences of the phenomena in the time interval $[0, t]$ can be regarded as a Poisson process.

Let us consider, as another example of an additive process, **Wiener process**. This process may be regarded as a continuous version of the random walk, and is defined as an additive process x_t, $0 \leq t < \infty$, for which the distribution Φ_{st} of $x_t - x_s$ is the normal distribution $N(\bullet; 0, t-s)$. Since this Φ_{st} also satisfies condition (2.1.2) of Theorem 2.1.2 (cf. §1.4), we can apply this theorem again to obtain such an additive process.

If x_t is an additive process, then so is the process $\alpha \cdot x_t + \beta \cdot t + \gamma$. Also, if x_t, $a \leq t < b$, and y_t, $a \leq t < b$, are both additive processes, and if the random vectors $\boldsymbol{x} = \prod_t x_t$ and $\boldsymbol{y} = \prod_t y_t$ are independent (which is the same as saying "for an arbitrary choice of $t_1, t_2, \cdots, t_n$, n-dimensional random vectors $\prod_i x_{t_i}$ and $\prod_i y_{t_i}$ are independent"), then $\alpha \cdot x_t + \beta \cdot y_t$, $a \leq t < b$, is also an additive process. This fact remains true even when more than two additive processes are involved. Utilizing this fact, we can construct, as we will see in §2.12, more general additive processes starting with Poisson processes and Wiener processes.

2.3. Inequalities Concerning Sums of Independent Random Variables

THEOREM 2.3.1 (Kolmogorov's Inequality). *Suppose $x_1, x_2, \cdots, x_n$ is a sequence of independent random variables, for which*

$$E(x_i) = 0, \quad V(x_i) < \infty, \quad i = 1, 2, \cdots, n. \tag{2.3.1}$$

Then

$$P\{\omega / \max_{k=1}^{n} |x_1 + \cdots + x_k| \geq c\} \leq \frac{1}{c^2} \sum_{k=1}^{n} V(x_k). \tag{2.3.2}$$

PROOF. Let

$$A_k = \{\omega / |x_1|, |x_1 + x_2|, \cdots, |x_1 + \cdots + x_{k-1}| < c, |x_1 + \cdots + x_k| \geq c\}, \quad 1 \leq k \leq n,$$

and denote by $a_k(\omega)$ the indicator function of the set A_k. $a_k(\omega)$ is measurable with respect to $x_1, x_2, \cdots, x_k$. From the definition of the A_k's it follows that the sets $A_1, A_2, \cdots, A_n$ are pairwise disjoint and the left-hand side of the inequality (2.3.2) equals

$$P(A_1 \cup A_2 \cup \cdots \cup A_n) = \sum_{k=1}^{n} P(A_k).$$

Since $E(x_i) = 0$, and since $\sum_{k=1}^{n} a_k \leq 1$, we have

$$V(x_1 + \cdots + x_n) = E((x_1 + \cdots + x_n)^2) \geq \sum_{k=1}^{n} E(a_k(x_1 + \cdots + x_n)^2).$$

Now let us look at the following identity:

$$\begin{aligned} E(a_k(x_1 + \cdots + x_n)^2) &= E(a_k(x_1 + \cdots + x_k)^2) \\ &\quad + 2E(a_k(x_1 + \cdots + x_k)(x_{k+1} + \cdots + x_n)) + E(a_k(x_{k+1} + \cdots + x_n)^2). \end{aligned}$$

The first term on the right-hand side of the identity above is greater than or equal to $c^2 P(A_k)$ due to the definition of a_k. Since $a_k(x_1 + \cdots + x_k)$ and $(x_{k+1} + \cdots + x_n)$ are measurable with respect to $(x_1, \cdots, x_k)$ and $(x_{k+1}, \cdots, x_n)$, respectively, they are independent random variables, and since $E(x_{k+1} + \cdots + x_n) = 0$, the second term on the right-hand side equals 0. It is clear that the third term on the right-hand side is greater than or equal to 0. Therefore, we obtain from the identity above that the left-hand side is greater than or equal to $c^2 P(A_k)$, from which it follows that

$$V(x_1 + \cdots + x_n) \geq c^2 \sum_{k=1}^{n} P(A_k).$$

Furthermore, since the left-hand side above equals $\sum_{k=1}^{n} V(x_k)$ because of the independence of $\{x_k\}$, we obtain, by putting together the observations made above, that the inequality (2.3.2) is valid. □

THEOREM 2.3.2 (Ottaviani's Inequality). *If $x_1, \cdots, x_n$ is a sequence of independent random variables for which*

$$P\{\omega / |x_{k+1} + \cdots + x_n| \leq c\} \geq 1/2, \quad k = 0, 1, \cdots, n-1, \tag{2.3.3}$$

then

$$P\{\omega / \max_{k=1}^{n} |x_1 + \cdots + x_k| > 2c\} \leq 2P\{\omega / |x_1 + \cdots + x_n| > c\}. \tag{2.3.4}$$

PROOF. Let

$$A_k = \{\omega / |x_1|, |x_1 + x_2|, \cdots, |x_1 + \cdots + x_{k-1}| \leq 2c, |x_1 + \cdots + x_k| > 2c\}, \ 1 \leq k \leq n,$$

$$B_k = \{\omega / |x_{k+1} + \cdots + x_n| \leq c\}, \ 1 \leq k \leq n-1.$$

Then $A_1, A_2, \cdots, A_n$ are pairwise disjoint, and the ω-set appearing on the left-hand side of the inequality (2.3.4) equals the union $A_1 \cup A_2 \cup \cdots \cup A_n$.

Since the inequalities $|x_1 + \cdots + x_k| > 2c$ and $|x_{k+1} + \cdots + x_n| \le c$ together imply that $|x_1 + \cdots + x_n| > c$ holds, we have, if we denote the ω-set appearing on the right-hand side of (2.3.4) by C,

$$(A_1 \cap B_1) \cup (A_2 \cap B_2) \cup \cdots \cup (A_n \cap B_n) \subseteq C \quad (\text{where } B_n = \Omega).$$

Since A_k is measurable with respect to $(x_1, \cdots, x_k)$, while B_k is measurable with respect to $(x_{k+1}, \cdots, x_n)$, and since $(x_1, \cdots, x_k)$ and $(x_{k+1}, \cdots, x_n)$ are independent, we get $P(A_k \cap B_k) = P(A_k)P(B_k)$, and as $P(B_k) \ge 1/2$ by hypothesis, we have $P(A_k \cap B_k) \ge P(A_k)/2$. Consequently,

$$\begin{aligned} P(C) &\ge P(A_1 \cap B_1) + P(A_2 \cap B_2) + \cdots + P(A_n \cap B_n) \\ &\ge \frac{1}{2}(P(A_1) + P(A_2) + \cdots + P(A_n)), \end{aligned}$$

from which it follows that $P(C) \ge \frac{1}{2}P(A_1 \cup A_2 \cup \cdots \cup A_n)$. □

2.4. 0-1 Law

An ω-set A is said to be measurable with respect to a random vector $\boldsymbol{x}(\omega) = \prod_{\lambda \in \Lambda} x_\lambda(\omega)$ (or with respect to x_λ, $\lambda \in \Lambda$), if the indicator function $a(\omega)$ of A is measurable with respect to $\boldsymbol{x}(\omega)$. We denote by $\boldsymbol{B}(\boldsymbol{x})$ or by $\boldsymbol{B}(x_\lambda, \lambda \in \Lambda)$ the smallest Borel field of subsets of Ω containing all sets of the form $\{\omega/x_\lambda \le \xi\}$, where λ is an arbitrary element of the set Λ and ξ is an arbitrary real number. The statement "A is measurable with respect to $\boldsymbol{x}(\omega)$" is equivalent to the fact that A is an element of $\boldsymbol{B}(\boldsymbol{x})$. It is also equivalent to the fact that A is of the form $\{\omega/\boldsymbol{x} \in E\}, E \in \boldsymbol{B}^\Lambda$.

For $\lambda_1, \lambda_2, \cdots, \lambda_n \in \Lambda$ and $E \in \boldsymbol{B}^n$, the set $\{\omega/(x_{\lambda_1}, x_{\lambda_2}, \cdots, x_{\lambda_n}) \in E\}$ is measurable with respect to $\boldsymbol{x}(\omega)$. Also if $\{\lambda_1, \lambda_2, \cdots\}$ is a sequence of elements of Λ, the set of all ω for which the sequence $\{x_{\lambda_n}(\omega), n = 1, 2, \cdots\}$ converges is a set measurable with respect to $\boldsymbol{x}$.

LEMMA 2.4.1. *Suppose a set A is measurable with respect to x_λ, $\lambda \in \Lambda$. Then, for every $\epsilon > 0$, there exists a set A_ϵ, measurable with respect to $x_{\lambda_1}, x_{\lambda_2}, \cdots, x_{\lambda_n}$, a finite set of elements suitably chosen from x_λ, $\lambda \in \Lambda$, and for which*

$$P(A - A_\epsilon) + P(A_\epsilon - A) < \epsilon \quad \textit{holds.} \tag{2.4.1}$$

PROOF. The family of all sets A satisfying the property stated in the lemma is easily seen to be a Borel field. Furthermore, it is obvious that every set of the form $\{\omega/x_\lambda \le \xi\}$ satisfies this property, and therefore, every set belonging to $\boldsymbol{B}(\boldsymbol{x})$ satisfies the same property. □

A family A_λ, $\lambda \in \Lambda$, of sets in $\boldsymbol{B}$ is said to be **independent**, if the corresponding family a_λ, $\lambda \in \Lambda$, of their indicator functions is a family of independent random variables.

LEMMA 2.4.2. *If $A_1, A_2, \cdots$ is a sequence of independent sets, then*

$$P(\bigcap_n A'_n) = \prod_n P(A'_n) \quad \textit{holds, where } A'_n = A_n \textit{ or } A_n^c. \tag{2.4.2}$$

LEMMA 2.4.3. *Suppose that $\boldsymbol{x}_\lambda$, $\lambda \in \Lambda$, is a family of independent random vectors, and suppose further that for each $\lambda \in \Lambda$, A_λ is a set measurable with respect to $\boldsymbol{x}_\lambda$. Then A_λ, $\lambda \in \Lambda$, is a family of independent measurable sets.*

THEOREM 2.4.1 (Kolmogorov's 0-1 Law). *Suppose $x_1, x_2, \cdots$ is a sequence of independent random variables, and suppose that a set A is measurable with respect to $x_n, x_{n+1}, \cdots$ for each n. Then*

$$P(A) = 0 \quad or \quad 1. \tag{2.4.3}$$

PROOF. Since A is measurable with respect to $x_1, x_2, \cdots$, for any given $\epsilon > 0$ we can choose n sufficiently large and a set A_ϵ, which is measurable with respect to $x_1, x_2, \cdots, x_n$, and which approximates the set A in the sense of (2.4.1). As A is measurable with respect to $x_{n+1}, x_{n+2}, \cdots$ as well, and since $\{x_n\}$ is independent, it follows from Lemma 2.4.3 that A and A_ϵ are independent. Consequently, we have $P(A \cap A_\epsilon) = P(A)P(A_\epsilon)$, and using (2.4.1), we obtain

$$|P(A) - P(A_\epsilon)| < \epsilon, \quad |P(A) - P(A \cap A_\epsilon)| < \epsilon.$$

By letting $\epsilon \to 0$, we then get $P(A) = P(A)^2$, from which it follows that $P(A) = 0$ or 1. □

By using Theorem 2.4.1, we can prove the following: If $x_1, x_2, \cdots$ is a sequence of independent random variables, then

$$P\{\omega / \textstyle\sum_n x_n \text{ converges}\} = 0 \quad \text{or} \quad 1,$$

$$P\{\omega / \tfrac{1}{n} \textstyle\sum_{\nu=1}^n x_\nu \to 0\} = 0 \quad \text{or} \quad 1.$$

THEOREM 2.4.2 (Borel-Cantelli's Theorem). *If a sequence $\{A_n\}$ of events satisfies the condition*

$$\sum P(A_n) < \infty, \tag{2.4.4}$$

then

$$P(\overline{\lim} A_n) = 0, \quad P(\underline{\lim} A_n^c) = 1. \tag{2.4.5}$$

If $\{A_n\}$ is independent, and if

$$\sum P(A_n) = \infty, \tag{2.4.6}$$

then

$$P(\overline{\lim} A_n) = 1, \quad P(\underline{\lim} A_n^c) = 0. \tag{2.4.7}$$

PROOF. If we suppose that condition (2.4.4) is satisfied, then

$$P(\overline{\lim} A_n) = P(\bigcap_k \bigcup_{n \geq k} A_n) \leq P(\bigcup_{n \geq k} A_n) \leq \sum_{n \geq k} P(A_n) \to 0,$$

which proves the first identity of (2.4.5). The second identity follows easily from the first since $\underline{\lim} A_n^c = (\overline{\lim} A_n)^c$.

Next, let us assume that $\{A_n\}$ is independent and that it satisfies (2.4.6). Then

$$P(\underline{\lim} A_n^c) = P(\bigcup_k \bigcap_{n \geq k} A_n^c) \leq \sum_k P(\bigcap_{n \geq k} A_n^c).$$

From Lemma 2.4.2 it follows that

$$P(\bigcap_{n \geq k} A_n^c) = \prod_{n \geq k} P(A_n^c) = \prod_{n \geq k} (1 - P(A_n)).$$

Since $\sum_{n \geq k} P(A_n) = \infty$ holds because of the assumption (2.4.6), the infinite product appearing on the right-hand side of the equality above equals 0, and therefore, $P(\underline{\lim} A_n^c) = 0$ as well. Thus, we conclude $P(\overline{\lim} A_n) = 1$. □

2.5. Convergence of Additive Sequences

Let x_n, $n = 0, 1, 2, \cdots$, be an additive sequence. Then, if we put $y_n = x_n - x_{n-1}$, $\{y_n\}$ becomes a sequence of independent random variables. In this section we will investigate conditions that guarantee the convergence of the sequence $\{x_n\}$. If we denote by C the set of all ω for which the sequence $x_n(\omega)$, $n = 0, 1, 2, \cdots$, converges, then

$$\begin{aligned} C &= \bigcap_p \bigcup_k \bigcap_{m,n>k} \{\omega / |x_m(\omega) - x_n(\omega)| < 1/p\} \\ &= \bigcap_p \bigcup_k \bigcap_{m,n>k} \{\omega / |y_{n+1}(\omega) + \cdots + y_m(\omega)| < 1/p\}, \end{aligned}$$

and therefore, the set C is measurable with respect to $y_1, y_2, \cdots$. Since the fact that $x_n(\omega)$, $n \geq 0$, converges is equivalent to the fact that $x_n(\omega) - x_N(\omega)$, $n \geq N$, converges for any fixed N, C can be written also as

$$C = \bigcap_p \bigcup_{k \geq N} \bigcap_{m,n>k} \{\omega / |y_{n+1}(\omega) + \cdots + y_m(\omega)| < 1/p\}.$$

Thus, we see that for each N, C is measurable with respect to $y_{N+1}, y_{N+2}, \cdots$. Therefore, we can conclude by applying Kolmogorov's 0-1 Law that

$$P(C) = 0 \text{ or } 1. \tag{2.5.1}$$

Next, we would like to see when $P(C) = 1$. Let us first give a sufficient condition for this to occur.

THEOREM 2.5.1. *If both of the sequences $\{E(x_n)\}$ and $\{V(x_n)\}$ converge, then $P(C) = 1$ holds; namely, $\{x_n\}$ converges a.e.*

PROOF. Since $x_n = (x_n - E(x_n)) + E(x_n)$ and $V(x_n) = V(x_n - E(x_n))$, we may assume without loss of generality that $E(x_n) = 0$ holds for each n. As

$$\sum_{n=1}^{\infty} V(y_n) = \lim_{n \to \infty} V(x_n) < \infty$$

holds, there exists, for any given $\epsilon > 0$, $n(\epsilon)$ such that whenever $n \geq n(\epsilon)$,

$$\sum_{k=n}^{\infty} V(y_k) < \epsilon \quad \text{holds.}$$

From Kolmogorov's Inequality it follows that

$$P\{\omega / \max_{1 \leq k \leq m} |y_{n+1} + \cdots + y_{n+k}| > a\} \leq \frac{\epsilon}{a^2}.$$

Letting $m \to \infty$, we get

$$P\{\omega / \sup_k |y_{n+1} + \cdots + y_{n+k}| > a\} \leq \frac{\epsilon}{a^2}.$$

Using the inequality

$$|y_{n+k} + \cdots + y_{n+\ell}| \leq |y_{n+1} + \cdots + y_{n+k-1}| + |y_{n+1} + \cdots + y_{n+\ell}| \quad (k < \ell),$$

we obtain from the result above that

$$P\{\omega / \sup_{\ell > k} |y_{n+k} + \cdots + y_{n+\ell}| > 2a\} \leq \frac{\epsilon}{a^2} \quad (n \geq n(\epsilon)),$$

and letting $n \to \infty$, we see that

$$P\{\omega / \lim_{n\to\infty} \left[\sup_{\ell>k} |y_{n+k} + \cdots + y_{n+\ell}|\right] > 2a\} \leq \frac{\epsilon}{a^2} \text{ holds.}$$

Letting $\epsilon \to 0$ first and then $a \to 0$, we obtain

$$P\{\omega / \lim_{n\to\infty} \left[\sup_{\ell>k} |y_{n+k} + \cdots + y_{n+\ell}|\right] > 0\} = 0,$$

which implies that the P-measure of the complement of the set C equals 0. □

The next theorem gives a necessary and sufficient condition for almost everywhere convergence of $\{x_n\}$.

THEOREM 2.5.2 (Three Series Theorem). *Suppose we define*

$$y'_n(\omega) = \begin{cases} y_n(\omega), & \textit{when } |y_n(\omega)| \leq 1, \\ 0, & \textit{when } |y_n(\omega)| > 1. \end{cases}$$

Then a necessary and sufficient condition for almost everywhere convergence of $\{x_n\}$ is given by the convergence of every one of the following three numerical series:

$$\sum E(y'_n), \quad \sum V(y'_n), \quad \sum P\{\omega / y_n \neq y'_n\}. \tag{2.5.2}$$

PROOF. (Sufficiency) Since the first two series converge, it follows from the preceding theorem that $\sum y'_n$ converges almost everywhere. From the convergence of the third series we can conclude by using Borel-Cantelli's theorem that, with probability 1, $y'_n(\omega) = y_n(\omega)$ holds for all but a finite number of n. Consequently, the almost everywhere convergence of y_n follows from the almost everywhere convergence of y'_n.

(Necessity) If $\sum P\{\omega / y_n \neq y'_n\} = \infty$, then it follows from Borel-Cantelli's theorem that, with probability 1, $y_n(\omega) \neq y'_n(\omega)$ occurs for infinitely many n. Since the fact that $y_n(\omega) \neq y'_n(\omega)$ is equivalent to the fact that $|y_n(\omega)| > 1$, this means that, with probability 1, $|y_n(\omega)| > 1$ occurs for infinitely many n, and therefore, $\sum y_n$ diverges with probability 1. Thus, in order for $\sum y_n$ and hence for x_n to converge almost everywhere, we must have $\sum P\{\omega / y_n \neq y'_n\} < \infty$. If this condition is satisfied, again by Borel-Cantelli's theorem we see that the almost everywhere convergence of $\sum y_n$ follows from that of $\sum y'_n$. Let us denote by Φ_n the distribution function of y'_n, by $\check{\Phi}_n$ the reflection of Φ_n, and by $\tilde{\Phi}_n$ the convolution $\Phi_n * \check{\Phi}_n$. Since the support of Φ_n is contained in $[-1, 1]$, the support of $\tilde{\Phi}_n$ is contained in $[-2, 2]$. It is obvious that

$$M(\tilde{\Phi}_n) = 0, \quad V(\tilde{\Phi}_n) = 2V(\Phi_n) = 2V(y'_n)$$

hold. If we let $\varphi_n(z)$ be the characteristic function of Φ_n, then the characteristic function of $\tilde{\Phi}_n$ is given by $|\varphi_n(z)|^2$. Since $\sum y'_n$ converges almost everywhere as we have proved above, we see that

$$\lim_{n\to\infty} E(e^{iz(y'_1+\cdots+y'_n)}) = E(e^{izx_\infty}), \quad x_\infty = \sum y'_n.$$

Hence, we conclude that $\prod_n \varphi_n(z)$ does not vanish in some neighborhood $|z| < a$ of 0, and therefore, $\prod_n |\varphi_n(z)|^2 > 0 \ (|z| < a)$. From this fact it follows that

$$\sum_n (1 - |\varphi_n(z)|^2) < \infty,$$

which means that

$$\sum_n \int (1 - \cos z\xi)\tilde{\Phi}_n(d\xi) < \infty.$$

Since $1 - \cos\xi > \frac{\xi^2}{3}$ holds for sufficiently small ξ, we see, by taking note of the fact that the support of $\tilde{\Phi}_n$ is contained in $[-2, 2]$, that for a sufficiently small positive number z

$$\sum_n \int \frac{z^2\xi^2}{3}\tilde{\Phi}_n(d\xi) < \infty \quad \text{holds.}$$

From this we conclude that $\sum V(\tilde{\Phi}_n)$ and hence $\sum V(y'_n)$ converges. By using Theorem 2.5.1, we deduce from this that $\sum_n (y'_n - E(y'_n))$ converges almost everywhere. As $\sum y'_n$ was already shown to converge almost everywhere, we finally conclude that $\sum_n E(y'_n)$ converges as well. □

THEOREM 2.5.3. *The following three conditions on an additive sequence x_n are mutually equivalent:*

(i) *The distribution of x_n converges.*
(ii) *x_n converges in probability.*
(iii) *x_n converges almost everywhere.*

PROOF. Implications (iii) ⇒ (ii) ⇒ (i) are obvious. Let us show the implication (i) ⇒ (ii). Let us denote by Φ_n the distribution of y_n and its characteristic function by φ_n. Then (i) implies that $\Phi_1 * \Phi_2 * \cdots * \Phi_n$ converges as $n \to \infty$ to some distribution Φ. Let us denote by φ the characteristic function of Φ. Then $|\varphi(z)| > b$ holds for some positive number b in some neighborhood $|z| \le a$ of $z = 0$. Furthermore, since $\varphi_1(z) \cdot \varphi_2(z) \cdots \varphi_n(z)$ converges to $\varphi(z)$ uniformly in every compact set of z, for any given $\epsilon > 0$ there exists $N(\epsilon)$ such that

$$m > n > N(\epsilon) \Rightarrow |\varphi_{n+1}(z) \cdot \varphi_{n+2}(z) \cdots \varphi_m(z) - 1| < \epsilon \quad (|z| \le a).$$

Suppose we set $\theta(z) = \varphi_{n+1}(z) \cdots \varphi_m(z)$, and denote by Θ the distribution corresponding to $\theta(z)$, namely the distribution of $x_m - x_n$. Then whenever $m > n > N(\epsilon)$ we have

$$|\theta(z) - 1| < \epsilon \quad (|z| \le a),$$

which means that

$$\left| \int (1 - e^{izx})\Theta(dx) \right| < \epsilon.$$

If we integrate the left-hand side above with respect to z from $-a$ to a and then divide by $2a$, we get

$$\int \left(1 - \frac{\sin xa}{xa}\right) \Theta(dx) < \epsilon.$$

Since there exists a constant $C > 0$ such that

$$1 - \frac{\sin x}{x} \ge C\frac{x^2}{1 + x^2},$$

we get from the above that

$$\int \frac{x^2a^2}{1 + x^2a^2}\Theta(dx) < \frac{\epsilon}{C},$$

from which it follows that

$$\int_{|x|\geq\eta}\Theta(dx) < \frac{(1+\eta^2a^2)\epsilon}{C\eta^2a^2},$$

and thus we conclude that whenever $m > n > N(\epsilon)$,

$$P\{\omega/|x_m - x_n| \geq \eta\} < \frac{(1+\eta^2a^2)\epsilon}{C\eta^2a^2}.$$

This implies that x_n converges in probability. Finally, to prove the implication (ii) $\Rightarrow$ (iii), we can argue as in the proof of Theorem 2.5.1 by using Ottaviani's Inequality instead of Kolmogorov's Inequality. □

2.6. Dispersion

As we mentioned before, the variance measures a degree of scattering of a distribution, but it does not necessarily exist, and therefore, it has limited applicability. So, let us introduce the notion of the dispersion, which exists for any distribution and plays the same role as the variance. The quantity defined for a 1-dimensional distribution Φ by

$$\delta(\Phi) = -\log\left[\int\int e^{-|x-y|}\Phi(dx)\Phi(dy)\right] \tag{2.6.1}$$

is called the **dispersion** of Φ. Also, we denote by $q(\Phi)$ the value of the double integral appearing inside of [ ] above, and call it the **degree of concentration** of Φ. For a random variable x, we also define $\delta(x)$ and $q(x)$ by $\delta(\Phi)$ and $q(\Phi)$, respectively, where Φ is the distribution of x. From the definitions it follows easily that the following facts are valid:

(i) $0 < q(\Phi) \leq 1, \quad 0 \leq \delta(\Phi) < \infty$,
(ii) $q(x) = q(x+a) = q(-x), \quad \delta(x) = \delta(x+a) = \delta(-x)$,
(iii) $q(\Phi) = 1 \Longleftrightarrow \delta(\Phi) = 0 \Longleftrightarrow \Phi$ is a δ-distribution,
(iv) $\delta(x_n) \to 0 \Longleftrightarrow$ "there exists a sequence of numbers $\{a_n\}$ such that $x_n - a_n$ converges in probability".

PROOF OF (iv). Proof of the implication $\Leftarrow$ is obvious. Let us prove the implication $\Rightarrow$. If $\delta(x_n) \to 0$, then

$$\int\int e^{-|x-y|}\Phi_n(dx)\Phi_n(dy) \to 1 \quad (\Phi_n \text{ is the distribution of } x_n).$$

Therefore, there exists a sequence $\{a_n\}$ such that

$$\int e^{-|x-a_n|}\Phi_n(dx) \to 1,$$

from which it follows that

$$\int(1 - e^{-|x-a_n|})\Phi_n(dx) \to 0,$$

and therefore, we get

$$\int_{|x-a_n|>\epsilon}\Phi_n(dx) \leq \frac{1}{1-e^{-\epsilon}}\int(1 - e^{-|x-a_n|})\Phi_n(dx) \to 0.$$ □

(v) $\delta(x_n) \to \infty \Longleftrightarrow Q_n(\ell) \equiv \sup_a \Phi_n[a-\ell, a+\ell] \to 0, \quad$ for every ℓ.

PROOF OF (v). Let us suppose that $\delta(x_n) \to \infty$; then $q(x_n) \to 0$. Since

$$\begin{aligned}\Phi_n[a-\ell, a+\ell]^2 &= \iint_{|x-a|,|y-a|\le\ell} \Phi_n(dx)\Phi_n(dy) \le \iint_{|x-y|\le 2\ell} \Phi_n(dx)\Phi_n(dy) \\ &\le e^{\ell} q(x_n),\end{aligned}$$

it follows that

$$Q_n(\ell) \le e^{\ell/2} q(x_n)^{1/2} \to 0.$$

Conversely, let us suppose that $Q_n(\ell) \to 0$ holds for every ℓ. Since

$$\begin{aligned}\int e^{-|x-y|}\Phi_n(dx) &\le \int_{|x-y|\ge\ell} e^{-|x-y|}\Phi_n(dx) \\ &\quad + \int_{|x-y|<\ell} e^{-|x-y|}\Phi_n(dx) \le e^{-\ell} + Q_n(\ell),\end{aligned}$$

we also have

$$q(x_n) = \iint e^{-|x-y|}\Phi_n(dx)\Phi_n(dy) \le e^{-\ell} + Q_n(\ell).$$

For any given $\epsilon > 0$, take ℓ so large that $e^{-\ell} < \frac{\epsilon}{2}$, and then choose n sufficiently large so that $Q_n(\ell) < \frac{\epsilon}{2}$. Then we get $q(x_n) < \epsilon$ for all sufficiently large n. Thus we conclude that $q(x_n) \to 0$, and therefore, $\delta(x_n) \to \infty$. □

(vi) Let $\varphi(z)$ be the characteristic function of Φ. Then

$$q(\Phi) = \frac{1}{\pi}\int \frac{|\varphi(z)|^2}{1+z^2}dz,$$

from which it follows that if $\Phi_n \to \Phi$, then $q(\Phi_n) \to q(\Phi)$, and hence $\delta(\Phi_n) \to \delta(\Phi)$ hold. Similarly, if $x_n \to x$ (convergence in probability), then $\delta(x_n) \to \delta(x)$ holds.

PROOF OF (vi). If we denote by Ψ the reflection of Φ, then the characteristic function of $\Phi * \Psi$ is given by $|\varphi(z)|^2$. Using this fact, we obtain

$$\begin{aligned}q(\Phi) &= \iint e^{-|x-y|}\Phi(dx)\Phi(dy) = \iint e^{-|x+y|}\Phi(dx)\Psi(dy) \\ &= \int e^{-|x|}(\Phi * \Psi)(dx) = \int \frac{1}{\pi}\int \frac{e^{izx}}{1+z^2}dz(\Phi * \Psi)(dx) \\ &= \frac{1}{\pi}\iint e^{izx}(\Phi * \Psi)(dx)\frac{dz}{1+z^2} = \frac{1}{\pi}\int \frac{|\varphi(z)|^2}{1+z^2}dz.\end{aligned}$$

□

(vii) **The principle of increase for the dispersion.** If x and y are independent random variables, then

$$q(x+y) \le q(x), \quad \text{and therefore,} \quad \delta(x+y) \ge \delta(x) \ \text{ holds.}$$

Furthermore, the equality holds in the above if and only if the distribution of y is a δ-distribution.

PROOF OF (vii). Let φ_1, φ_2 be the characteristic functions of x, y, respectively. We then have

$$q(x+y) = \frac{1}{\pi}\int \frac{|\varphi_1(z)\varphi_2(z)|^2}{1+z^2}dz \le \frac{1}{\pi}\int \frac{|\varphi_1(z)|^2}{1+z^2}dz = q(x).$$

In order for the equality to hold above, we must have $|\varphi_2(z)|^2 = 1$, whenever $|\varphi_1(z)| > 0$ holds. But, since $|\varphi_1(z)| > 0$ in some neighborhood of $z=0$, $|\varphi_2(z)|^2 = 1$

must hold throughout this neighborhood. Let Φ be the distribution of y, and let $\tilde{\Phi}$ be the convolution of Φ and its reflection $\check{\Phi}$. Then, the characteristic function of $\tilde{\Phi}$ is given by $|\varphi_2(z)|^2$, and furthermore, $\tilde{\Phi}$ is a symmetric distribution, and thus we have

$$\int \cos z\xi \tilde{\Phi}(d\xi) = 1, \quad \text{namely,} \quad \int (1 - \cos z\xi)\tilde{\Phi}(d\xi) = 0$$

holds in some neighborhood $|z| \leq a$ of $z = 0$. If we integrate with respect to z from $-a$ to a, and then divide by $2a$, we get

$$\int \left(1 - \frac{\sin a\xi}{a\xi}\right) \tilde{\Phi}(d\xi) = 0.$$

Since for some positive constant C, the inequality

$$1 - \frac{\sin z}{z} > C\frac{z^2}{1+z^2}$$

holds, we get

$$\int \frac{a^2\xi^2}{1+a^2\xi^2}\tilde{\Phi}(d\xi) \leq 0.$$

We thus conclude that $\tilde{\Phi}$ must be the unit distribution $\delta(\bullet; 0)$, and hence Φ is a δ-distribution. □

Let x_n be an additive sequence, and let $y_n = x_n - x_{n-1}$. Then x_{n+1} can be represented as the sum of independent random variables x_n and y_{n+1}. Therefore, we have $q(x_n) \geq q(x_{n+1})$ and $\delta(x_n) \leq \delta(x_{n+1})$. If we set $q = \lim q(x_n), \delta = \lim \delta(x_n)$, then we can show that the following theorem holds.

THEOREM 2.6.1. $\delta < \infty \Longleftrightarrow q > 0 \Longleftrightarrow$ *"There exists a sequence $\{a_n\}$ such that $\{x_n - a_n\}$ converges almost everywhere, and furthermore, $\delta = \delta(\lim_n(x_n - a_n))$ holds".*

PROOF. Let us suppose that $\delta < \infty$ and hence $q > 0$. We see that

$$q = \lim_n \frac{1}{\pi}\int \frac{|\varphi_1(z)|^2 \cdots |\varphi_n(z)|^2}{1+z^2}dz = \frac{1}{\pi}\int \prod_{n=1}^{\infty} |\varphi_n(z)|^2 \frac{dz}{1+z^2}.$$

Since $q > 0$, we must have $\prod_n |\varphi_n(z)|^2 > 0$ and hence

$$\sum_n [1 - |\varphi_n(z)|^2] < \infty$$

must hold on some set A of positive measure. Therefore, if we choose a suitable subset A_1 of positive measure of the set A, then we have that

$$\sum_n [1 - |\varphi_n(z)|^2] < C < \infty$$

holds on A_1 for some constant $C > 0$. We may assume that the Lebesgue measure $|A_1|$ of A_1 is finite. Let Φ_n be the distribution of y_n, and let $\tilde{\Phi}_n$ be the convolution of Φ_n and its reflection $\check{\Phi}_n$. Then we have

$$\sum_n \int (1 - \cos z\xi)\tilde{\Phi}_n(d\xi) < C. \tag{2.6.2}$$

Since $0 < |A_1| < \infty$, the function

$$f(\xi) \equiv \frac{1+\xi^2}{\xi^2} \int_{A_1} (1 - \cos z\xi) dz$$

is continuous on $0 < |\xi| < \infty$, and satisfies $f(\xi) > 0$. Furthermore, by the Riemann-Lebesgue Theorem, we get $f(\xi) \to |A_1| > 0$ as $|\xi| \to \infty$. We see also that when $|\xi| \to 0$,

$$f(\xi) \to \frac{1}{2} \int_{A_1} z^2 dz > 0.$$

Therefore, $f(\xi)$ has a strictly positive lower bound on R^1, and

$$\int_{A_1} (1 - \cos z\xi) dz > k \frac{\xi^2}{1+\xi^2} \quad \text{for some constant } k > 0.$$

Thus we can integrate the left-hand side of (2.6.2) term by term with respect to z on the set A_1 to get

$$\sum_n \int \frac{\xi^2}{1+\xi^2} \tilde{\Phi}_n(d\xi) < C' \quad \text{for some suitable constant } C' > 0.$$

Since

$$\int \frac{\xi^2}{1+\xi^2} \tilde{\Phi}_n(d\xi) = \iint \frac{(\xi - \eta)^2}{1 + (\xi - \eta)^2} \Phi_n(d\xi) \Phi_n(d\eta),$$

we can choose constants $\{b_n\}$ suitably to get

$$\sum_n \int \frac{(\xi - b_n)^2}{1 + (\xi - b_n)^2} \Phi_n(d\xi) < \sum_n \int \frac{\xi^2}{1+\xi^2} \tilde{\Phi}_n(d\xi) < C',$$

from which we can conclude that

$$\sum_n \int_{|\xi - b_n| > 1} \Phi_n(d\xi) < \infty, \quad \sum_n \int_{|\xi - b_n| \le 1} \xi^2 \Phi_n(d\xi) < \infty.$$

Suppose we set

$$z_n = \begin{cases} y_n - b_n, & \text{if } |y_n - b_n| \le 1, \\ 0, & \text{if } |y_n - b_n| > 1. \end{cases}$$

Then it follows from the results above that we have

$$\sum_n P\{\omega / z_n \ne y_n - b_n\} < \infty,$$

$$\sum_n E(z_n^2) < \infty.$$

By Borel-Cantelli's Theorem, we conclude from the first of these conditions that, with probability 1, $z_n = y_n - b_n$ holds for all but a finite number of n. From the second condition it follows that

$$\sum_n V(z_n) \le \sum_n E(z_n^2) < \infty,$$

which implies, in view of Theorem 2.5.1, that $\sum_n (z_n - c_n)$ with $c_n = E(z_n)$ converges almost everywhere, and therefore, $\sum_n (y_n - d_n)$ with $d_n = b_n + c_n$ converges almost everywhere also. By setting $a_n = d_1 + d_2 + \cdots + d_n$, we conclude that

$\{x_n - a_n\}$ also converges almost everywhere. If we let $\varphi(z)$ be the characteristic function of $\lim(x_n - a_n)$, then $|\varphi(z)|^2 = \prod_{n=1}^{\infty} |\varphi_n(z)|^2$, so we get $q(\lim(x_n - a_n)) = q$ and therefore, $\delta(\lim(x_n - a_n)) = \delta$. □

THEOREM 2.6.2. $\delta = \infty \iff q = 0 \iff Q_n(\ell) \equiv \sup_a P\{\omega/a - \ell < x_n \leq a + \ell\} \to 0$.

Proof of this theorem was already given in (v) above.

The sequence $\{a_n\}$ appearing in Theorem 2.6.1 above is called a **sequence of stabilizing constants**. If $\{a_n\}$ is a sequence of stabilizing constants and if $\{b_n\}$ is a convergent sequence, then $\{a_n + b_n\}$ is also a sequence of stabilizing constants. The following theorem tells us how to find a sequence of stabilizing constants.

THEOREM 2.6.3. *Suppose that the condition of Theorem* 2.6.1 *is satisfied. Then if we determine* c_n *so as to satisfy*

$$E(\arctan(x_n - c_n)) = 0,$$

then the sequence $\{c_n\}$ *becomes a sequence of stabilizing constants. This sequence is called the* **sequence of stabilizing constants of Doob**.

PROOF. Let us first show that c_n is determined uniquely by the requirement made above. If we let c_n change from $-\infty$ to ∞, then $E(\arctan(x_n - c_n))$ decreases strictly, monotonically and continuously from $\frac{\pi}{2}$ to $-\frac{\pi}{2}$. Therefore, there is one and only one c_n for which $E(\arctan(x_n - c_n)) = 0$. Now suppose that the condition of Theorem 2.6.1 is satisfied. Then there exists a sequence $\{a_n\}$ of stabilizing constants. So, Theorem 2.6.3 will be proved if we can show that the sequence $\{c_n - a_n\}$ is convergent. For that it suffices to show that there is only one limit point for the sequence $\{c_n - a_n\}$. So, suppose for some subsequence $p(n)$ of natural numbers that

$$c_{p(n)} - a_{p(n)} \to c.$$

Since by hypothesis $\{x_n - a_n\}$ converges almost everywhere, we denote by x its limit. Then

$$x_{p(n)} - c_{p(n)} = (x_{p(n)} - a_{p(n)}) + (a_{p(n)} - c_{p(n)}) \to x - c$$

and therefore, we get

$$E(\arctan(x - c)) = \lim_n E(\arctan(x_{p(n)} - c_{p(n)})) = 0,$$

from which it is easy to see that the limit point c of the sequence $\{c_n - a_n\}$ is unique. □

REMARK. When the condition of Theorem 2.6.1 is satisfied, then the series $\sum y_n$ is said to be of **convergence type** and when the condition of Theorem 2.6.2 is satisfied it is said to be of **divergence type**.

2.7. Simple Properties of Additive Processes

Let x_t, $t \in T = [a, b)$, be an additive process. From the definition of additive process and from the principle of the increase of the dispersion, it follows that

$$(s, t) \subseteq (u, v) \Rightarrow \delta(x_v - x_u) \geq \delta(x_t - x_s).$$

In particular, by setting $u = s = a$, we see that $\delta(t) \equiv \delta(x_t)$ is an increasing function of t. The set D of the points of discontinuity of $\delta(t)$ is at most countable, and is the disjoint union of the following three sets:

$$D^+ = \{t/\delta(t+0) > \delta(t), \delta(t-0) = \delta(t)\},$$
$$D^- = \{t/\delta(t+0) = \delta(t), \delta(t-0) < \delta(t)\},$$
$$D^0 = \{t/\delta(t+0) > \delta(t), \delta(t-0) < \delta(t)\}.$$

As we saw in the preceding section, for each t a constant $f(t)$ is uniquely determined by the condition

$$E\{\arctan(x_t - f(t))\} = 0.$$

Let us set

$$z_t = x_t - f(t)$$

so that $x_t = z_t + f(t)$.

Suppose $t_1 < t_2 < \cdots \to t$. Then z_{t_n}, $n = 1, 2, \cdots$, is an additive sequence, and as $\delta(t_n) \leq \delta(t)$, $\{z_{t_n} - c_n\}$ converges almost everywhere for a sequence of stabilizing constants $\{c_n\}$. Since $E(\arctan z_{t_n}) = 0$, we can take $c_n = 0$ for each n. This means that $\{z_{t_n}\}$ converges almost everywhere. Let us denote its limit by z_{t-}. z_{t-} is determined when t is fixed, except on a set of P-measure 0. (This exceptional set of P-measure 0 depends on t.) For another increasing sequence $t'_1 < t'_2 < \cdots \to t$, we can consider the same kind of limit, which we call z'_{t-}. If we put together $\{t_n\}$ and $\{t'_n\}$ and arrange them in increasing order to get a new sequence $t''_1 < t''_2 < \cdots \to t$ and let z''_{t-}, then we get that

$$z_{t-} = z''_{t-} = z'_{t-}$$

holds almost everywhere. Therefore, with probability 1, z_{t-} is determined independently of the choice of increasing sequence $\{t_n\}$. The exceptional set of ω does depend on t. In order to define z_{t+}, we let $t_n \downarrow t$. Then $\{z_{t_n} - z_{t_1}\}$ becomes an additive sequence, and since $\delta(z_{t_n} - z_{t_1}) = \delta(z_{t_1} - z_{t_n}) \leq \delta(z_{t_1} - z_t)$, the sequence $\{z_{t_n} - z_{t_1} - c_n\}$ converges almost everywhere, and hence $\{z_{t_n} - c_n\}$ converges almost everywhere. Since $E(\arctan(z_{t_n})) = 0$, it follows that $\{z_{t_n}\}$ converges almost everywhere. We set this limit to be z_{t+}. This is also determined independently of the choice of decreasing sequence $\{t_n\}$.

THEOREM 2.7.1.

$$t \notin D \iff P\{\omega/z_{t+} = z_t = z_{t-}\} = 1,$$
$$t \in D^+ \iff P\{\omega/z_{t-} = z_t\} = 1, \quad \delta(z_{t+} - z_t) > 0,$$
$$t \in D^- \iff P\{\omega/z_{t+} = z_t\} = 1, \quad \delta(z_t - z_{t-}) > 0,$$
$$t \in D^0 \iff \delta(z_{t+} - z_t) > 0, \quad \delta(z_t - z_{t-}) > 0.$$

Before we start the proof of the theorem, we need the following lemma.

LEMMA 2.7.1. *Suppose for each n, random variables $\{x_n, y_n, \cdots, u_n\}$ are independent, and suppose $x_n, y_n, \cdots, u_n$ converge almost everywhere to $x, y, \cdots, u$, respectively, as $n \to \infty$. Then $x, y, \cdots, u$ are also independent.*

PROOF.

$$\begin{aligned} E\{\exp\{i(\theta x + \varphi y + \cdots + \psi u)\}\} &= \lim_{n\to\infty} E\{\exp\{i(\theta x_n + \varphi y_n + \cdots + \psi u_n\}\} \\ &= \lim_{n\to\infty} E(e^{i\theta x_n})E(e^{i\varphi y_n})\cdots E(e^{i\psi u_n}) \\ &= E(e^{i\theta x})E(e^{i\varphi y})\cdots E(e^{i\psi u}). \end{aligned}$$

If we denote by $\Theta, \Phi, \cdots, \Psi$ the distribution function of $x, y, \cdots, u$, respectively, and let the distribution of the random vector $(x, y, \cdots, u)$ be F, then from the equation above, we see that the characteristic function of F coincides with the characteristic function of $\Theta \times \Phi \times \cdots \times \Psi$, and therefore, the two distributions must be equal. This means that $x, y, \cdots, u$ are independent. □

PROOF OF THEOREM 2.7.1. From Lemma 2.7.1 above it follows that $z_t - z_{t-}$ and z_{t-} are independent. Since $z_t = (z_t - z_{t-}) + z_{t-}$, and since $\delta(z_t) = \lim_{s\uparrow t}\delta(z_s) = \delta(t-0)$, we conclude that, if $\delta(t-0) = \delta(t)$ holds, then, with probability 1, $z_t - z_{t-}$ is a constant. This constant must equal 0, since $E(\arctan z_t) = 0$, and $E(\arctan z_{t-}) = \lim_{s\uparrow t} E(\arctan z_s) = 0$. Thus we have $P(z_t = z_{t-}) = 1$. If, on the other hand, $\delta(t-0) < \delta(t)$, then $z_t - z_{t-}$ is not a constant, and $\delta(z_t - z_{t-}) > 0$. We get similar conclusions for $z_{t+} - z_t$. Putting these results together, we obtain the assertion of Theorem 2.7.1. □

Let us next try to determine an additive process u_t by collecting only the jumps $z_{t+} - z_{t-}$, $t \in D$, of z_t. However, since there is a matter of convergence of infinite sums to worry about, we usually cannot adopt a naive idea such as defining u_t simply as

$$u_t = \sum_{\substack{s\le t\\ s\in D}} (z_{t+} - z_{t-}).$$

So, let us enumerate the elements of the countable set D as $s_1, s_2, \cdots$, and then consider for each n,

$$u_t^{(n)} = \sum_{\substack{1\le i\le n\\ s_i\le t}} (z_{s_i+} - z_{s_i-}) - c_t^{(n)},$$

where $c_t^{(n)}$ is a constant chosen so that $E(\arctan u_t^{(n)}) = 0$, and furthermore, if $s_i = t$, then we take z_{s_i+} to be equal to z_t. $u_t^{(n)}$ is a partial sum of a countably infinite sum of independent random variables, and since, by the principle of increase of dispersion, $\delta(u_t^{(n)}) \le \delta(z_t)$, $u_t^{(n)}$ converges a.e. Let us define this limit as u_t. Then, it is clear that $E(\arctan u_t) = 0$, and noting this fact, we can also conclude that the same u_t will result even if we change the enumeration of the set D. From this definition the following theorem is obvious.

THEOREM 2.7.2. *u_t is an additive process, and $\delta(u_t)$ is a purely discontinuous increasing function, which increases only by jumps at the points of D. Furthermore,*

$$\begin{aligned} t \notin D \Rightarrow & P\{\omega/u_{t-} = u_t = u_{t+}\} = 1, \\ t \in D \Rightarrow & P\{\omega/u_{t+} - u_t = z_{t+} - z_t + \textit{constant}\} = 1 \quad \textit{and} \\ & P\{\omega/u_t - u_{t-} = z_t - z_{t-} + \textit{constant}\} = 1. \end{aligned}$$

Next, set $v_t = z_t - u_t - c_t$, where c_t is a constant chosen to make $E(\arctan v_t) = 0$, and let us investigate the properties of v_t.

THEOREM 2.7.3. *v_t is an additive process, and $\delta(v_t)$ is a continuous function of t. Furthermore,*

$$P\{\omega/v_{t-} = v_t = v_{t+}\} = 1 \tag{2.7.1}$$

holds.

PROOF. Let $v_t^{(n)} = z_t - u_t^{(n)} - c_t$, where $u_t^{(n)}$ is the finite sum of independent random variables which we considered above when we defined u_t. Then, clearly, $v_t^{(n)}$ is an additive process and converges a.e. to v_t. Therefore, by using Lemma 2.7.1, we can conclude that v_t is also an additive process. From $E(\arctan(v_t)) = 0$, we see that v_{t-}, v_{t+} are determined, and hence so are c_{t-0}, c_{t+0}. Theorem 2.7.2 implies that, with probability 1, both $v_{t+} - v_t$ and $v_t - v_{t-}$ are constants. Since $E(\arctan v_t) = 0$, we see that both of these constants are 0. Therefore, we see that (2.7.1) holds, and by using Theorem 2.7.1, we can conclude also that $\delta(v_t)$ has no points of discontinuity. □

Summarizing what we have obtained above, we see that we have the decomposition

$$x_t = u_t + v_t + g(t), \quad g(t) = f(t) + c_t,$$

where $g(t)$ is a function only of t, u_t, v_t are both additive processes, $\delta(u_t)$ is purely discontinuous, and $\delta(v_t)$ is continuous. Furthermore, $E(\arctan u_t) = E(\arctan v_t) = 0$ holds, and hence (2.7.1) is valid. A relation between u_t and v_t is explained by the following theorem.

THEOREM 2.7.4. *The two additive processes u_t, $t \in T$, and v_t, $t \in T$, are independent (meaning that the two random vectors $\prod_t u_t$ and $\prod_t v_t$ are independent).*

PROOF. Using the fact that $z_{s_i-} - z_{s_{i-1}}, z_{s_i} - z_{s_i-}, z_{s_i+} - z_{s_i}$, $i = 1, 2, \cdots, n$, are independent and using the properties concerning independence discussed in §1.3, we can conclude that for each n, $u_t^{(n)}$, $t \in T$, and $v_t^{(n)}$, $t \in T$, are independent random variables. Therefore, for arbitrary $t_1, t_2, \cdots, t_m$, m-dimensional random vectors $\prod_i u_{t_i}^{(n)}$ and $\prod_i v_{t_i}^{(n)}$ are independent. Since Lemma 2.7.1 is valid also for the case of random vectors, we can conclude by taking the limit as $n \to \infty$ that $\prod_i u_{t_i}$ and $\prod_i v_{t_i}$ are independent, and thus, $\prod_t u_t$ and $\prod_t v_t$ are independent. □

From what we described above it is clear that in order to study additive processes, it is sufficient to investigate separately those for which $\delta(x_t)$ are purely discontinuous, and those for which $\delta(x_t)$ are continuous. Additive processes of the former type are always constructed in the following way. Fix a countable subset D of $[a, b)$, and for each $s \in D$ associate two random variables ξ_s and η_s in such a way that ξ_s, η_s, $s \in D$, give a family of independent random variables. Furthermore, assume that for every t with $a \leq t < b$,

$$\xi_t = \sum_{\substack{s<t \\ s \in D}} (\xi_s + \eta_s)$$

is of the convergent type in the sense we defined at the end of §2.6. If we subtract terms of the sequence of stabilizing constants of Doob, and then calculate the convergent sum, calling the summand x_t, we get an additive process with purely discontinuous $\delta(x_t)$.

We will explain in the succeeding sections the structure of x_t for which $\delta(x_t)$ is continuous.

2.8. Separability of Stochastic Processes

Let $x_t(\omega)$, $t \in T$, be an arbitrary stochastic process. For each individual t, $x_t(\omega)$ is a measurable function of ω, and therefore, the sets

$$\begin{aligned} A &= \{\omega / x_{t_1} \leq a_1, x_{t_2} \leq a_2, \cdots, x_{t_n} \leq a_n\}, \\ B &= \{\omega / \overline{\lim_n}\, x_{t_n}(\omega) \leq a\}, \\ C &= \{\omega / x_t(\omega) \leq 1 \text{ for every rational number } t \in T\} \end{aligned}$$

are all measurable sets and we can find the P-measure of these sets. Furthermore, if for every $t \in T$

$$P(C_t) = 1, \quad \text{where } C_t = \{\omega / x_t(\omega) \leq 1\},$$

then C can be written as

$$C = \bigcap_{\substack{t \in T \\ t \text{ is rational}}} C_t$$

so that C is an intersection of a countable number of sets, each of probability 1, and therefore, we have $P(C) = 1$. However, if we set, instead of C,

$$C' = \{\omega / x_t(\omega) \leq 1 \text{ for every } t \in T\},$$

then $C' = \bigcap_{t \in T} C_t$ so that C' is an intersection of an uncountable number of sets, each of P-measure 1, and we cannot decide in general whether such an intersection is measurable. Sometimes, such a set may be measurable and may even have P-measure 1, or may be measurable with P-measure 0, or may be non-measurable. From this example we see that we must emphasize the following important fact:

"According to the definition of stochastic processes it is a simple matter to discuss the probability of events which concern at most a countable number of time points (t), but special care must be taken when we discuss the probability of events involving the values of x_t at an uncountable number of time points."

In this connection, we note that there are some important conditions on a stochastic process x_t, $t \in T$, which depend on an uncountable number of time points such as

(i) $x_t(\omega)$ is a continuous function of t,
(ii) $x_t(\omega)$ is a bounded function of t,
(iii) $x_t(\omega)$ is an increasing function of t.

In order to discuss probabilities of these conditions, it becomes necessary to put some restrictions on the stochastic processes we deal with. Doob is the first person to realize the importance of this point, and he introduced the restriction called **separability** to stochastic processes, and succeeded in overcoming this difficulty.

DEFINITION 2.8.1. We say that a stochastic process x_t, $t \in T$, is separable if there exists a countable subset S of T such that the following condition (S) is satisfied:

$$(S) \quad P\{\omega/\text{for every } t \in T, \varliminf_{\substack{s \to t \\ s \in S}} x_s(\omega) \leq x_t \leq \varlimsup_{\substack{s \to t \\ s \in S}} x_s(\omega) \text{ holds}\} = 1.$$

For a separable stochastic process $x_t(\omega)$, $t \in T$, all of the events (i), (ii), and (iii) described above are measurable. Also, a process x_t, $t \in T$, for which the probability of the event (i) is 1 is separable.

We say that two stochastic processes x_t, $t \in T$, and y_t, $t \in T$, are **equivalent in the weak sense** if

$$P\{\omega/x_t(\omega) = y_t(\omega)\} = 1 \quad \text{for every } t \in T.$$

From this it follows that for arbitrarily chosen $t_1, t_2, \cdots, t_n \in T$ and for an arbitrary $E_n \in \boldsymbol{B}^n$,

$$P\{\omega/(x_{t_1}(\omega), \cdots, x_{t_n}(\omega)) \in E_n\} = P\{\omega/(y_{t_1}(\omega), \cdots, y_{t_n}(\omega)) \in E_n\}$$

holds. More generally, for $E \in \boldsymbol{B}^T$,

$$P\{\omega/\prod_t x_t \in E\} = P\{\omega/\prod_t y_t \in E\}$$

holds also. However, since the set $\boldsymbol{C}$ of all continuous functions defined on T, as a subset of R^T, does not belong to $\boldsymbol{B}^T$, the equality

$$P\{\omega/\prod_t x_t \in \boldsymbol{C}\} = P\{\omega/\prod_t y_t \in \boldsymbol{C}\}$$

does not necessarily hold. If x_t, $t \in T$, is not separable, then the ω-set appearing on the left-hand side above may not even be measurable. The same can be said for y_t. Even when one of the two stochastic processes is separable, and the two are equivalent in the weak sense, we cannot conclude in general that the other is also separable.

According to Doob, we have the following:

THEOREM 2.8.1. *For an arbitrarily given stochastic process, there exists a separable stochastic process, which is equivalent in the weak sense to the given one. This is called a* **separable modification** *of the given process.*

In view of this theorem, we can see that it is enough to investigate only separable stochastic processes by considering separable modifications, if necessary.

2.9. Separable Poisson Processes

As we defined before, an additive process x_t, $t \in T$, is called a **Poisson process** if $x_t - x_s$ $(t > s)$ has the Poisson distribution $P(\bullet; \lambda(t-s))$. If it is separable, it is called a **separable Poisson process**. Since there exists a Poisson process, we see, by taking its separable modification, that a separable Poisson process exists also.

THEOREM 2.9.1. *If x_t, $t \in T = [a, b)$, is a separable Poisson process, then with probability* 1, *its sample process is a step function which increases only by jumps of magnitude* 1. (*The value at the jump point lies between the left and the right limits at that point.*)

PROOF. The definition of the separability states that there exists a countable subset S of T such that the following ω-set has P-measure 1:

$$\Omega' = \{\omega/ \varliminf_{\substack{s \to t \\ s \in S}} x_s(\omega) \le x_t \le \varlimsup_{\substack{s \to t \\ s \in S}} x_s(\omega), t \in T\}.$$

We note that when we fix t the set

$$\Omega_t = \{\omega/x_t(\omega) = \text{non-negative integer}\}$$

has P-measure 1. This is true since the distribution of x_t is $P(\bullet; \lambda(t-a))$. Also, since the distribution of $x_t - x_u$ $(t > u)$ is $P(\bullet; \lambda(t-u))$, P-measure of the set

$$\Omega_{ut} = \{\omega / x_t(\omega) - x_u(\omega) \geq 0\}$$

is 1 as well. As the set S is countable, if we define

$$\Omega'' = \bigcap_{s \in S} \Omega_s \cap \bigcap_{\substack{s<t \\ s,t \in S}} \Omega_{st} \cap \Omega',$$

then the set Ω'' also has P-measure 1. Now, if $\omega \in \Omega''$, then $x_t(\omega)$ as a function of t is a step function increasing by jumps of magnitude $1, 2, \cdots$.

The proof of this theorem will be complete if we can show that the P-measure of the set N equals 0, where N is the set of ω for which $x_t(\omega)$ has jumps of magnitude 2 or bigger. For this purpose it is enough to show that for each t_0 the set $N(t_0)$ has P-measure 0, where $N(t_0)$ is the set of ω for which $x_t(\omega)$ has jumps of magnitude 2 or bigger in the time interval $a \leq t \leq t_0$. So, let us define

$$E_{nk} = \left\{\omega \Big/ x\left(a + \frac{k}{n}(t_0 - a), \omega\right) - x\left(a + \frac{k-1}{n}(t_0 - a), \omega\right) \geq 2\right\}, \quad 1 \leq k \leq n.$$

Then, it holds that

$$N(t_0) \subset \bigcup_k E_{nk},$$

and therefore,

$$P(N(t_0)) \leq \sum_{k=1}^{n} P(E_{nk}) = O\left(n\left(\frac{1}{n}\right)^2\right) \to 0 \quad (n \to \infty). \qquad \square$$

DEFINITION 2.9.1. If for an arbitrary stochastic process x_t, $t \in T$,

$$P\{\omega / |x_t - x_s| > \epsilon\} \to 0 \quad (t \to s)$$

holds for every ϵ, then this stochastic process is said to be **continuous in probability** at s. x_t, $t \in T$, is said to be a **stochastic process continuous in probability** if it is continuous in probability at every point of T.

For a (separable) Poisson process considered above

$$P\{\omega / |x_t - x_s| > \epsilon\} = O(|t - s|) \to 0 \quad (t \to s)$$

holds, so it is continuous in probability. However, a sample process of a separable Poisson process is typically a step function, which is very far from a continuous function. Roughly speaking this means that each individual sample process has points of discontinuity, but where discontinuous time points are located varies with ω, and if we fix a time point the probability that discontinuity occurs at that point is 0. In fact, if we let N_t be the set of ω's for which $x_{t+0} = x_{t-0}$, then $P(N_t) = 1$, but $P(\bigcap_t N_t) = 0$.

If a stochastic process x_t is an additive process which is continuous in probability and the distribution of $x_t - x_s$ for each pair $(t > s)$ is a Poisson distribution, then x_t is called a **Poisson process in a wide sense**. If in this case we write $\lambda(t) = E(x_t)$, then since x_t is continuous in probability, $\lambda(t)$ becomes a continuous function of t. The distribution of $x_t - x_s$ then becomes $P(\bullet; \lambda(t) - \lambda(s))$. Poisson process discussed above corresponds to the case where $\lambda(t) = \lambda \cdot (t - a)$. In this sense, it is sometimes called a **temporally homogeneous** Poisson process.

THEOREM 2.9.2. *The same conclusion as for Theorem 2.9.1 above holds for a separable Poisson process in a wide sense as well.*

Conversely, we have the following:

THEOREM 2.9.3. *If an additive process which is continuous in probability has the property that, with probability 1, its sample process is a step function which increases only with jumps of magnitude 1, then it is a separable Poisson process in a wide sense.*

PROOF. Let x_t, $t \in T = [a, b)$, be the additive process satisfying the hypothesis of the theorem. From the assumption that its sample functions are step functions it follows easily that the process is separable. Therefore, it is enough to show that the distribution of $x_t - x_s \quad (t > s)$ is a Poisson distribution. From the assumption of continuity in probability we see that for $u \in T$ and $\epsilon > 0$ there exists a $\delta = \delta(\epsilon, u)$ such that whenever $|v - u| < \delta$,

$$P\{\omega / |x_v - x_u| > \epsilon\} < \epsilon$$

holds. Because of the Borel Covering Theorem, on a closed interval $s \leq u \leq t$, $\delta(\epsilon, u)$ can be chosen independently of u. Let us subdivide the closed interval $[s, t]$ into n equal parts, and let $y_{n1}, y_{n2}, \cdots, y_{nn}$ be the increments of x_t over the subintervals respectively, and let $y = y_{n1} + y_{n2} + \cdots + y_{nn} = x_t - x_s$. Define

$$y'_{nk} = \begin{cases} y_{nk}, & y_{nk} \leq 1, \\ 0, & y_{nk} \geq 2, \end{cases}$$

and let

$$y'_n = y'_{n1} + y'_{n2} + \cdots + y'_{nn}.$$

From the assumption on sample processes it follows that

$$P(y'_n \to y) = 1.$$

If we denote $p_{nk} = P\{\omega / y'_{nk} = 1\}$, then $p_{nk} = P\{\omega / y_{nk} = 1\}$ holds, and from what we observed above it follows that, as $n \to \infty$, $p_{nk} \to 0$ holds uniformly on k. We note also that $\{y'_{nk},\ k = 1, 2, \cdots, n\}$ are independent since $\{y_{nk},\ k = 1, 2, \cdots, n\}$ are.

Now we consider several possibilities:

(i) The case when $\sum_k p_{nk} \to \lambda \ (< \infty)$.

$$\begin{aligned} E(e^{izy}) &= \lim_n E(e^{izy'_n}) = \lim_n \prod_k E(e^{izy'_{nk}}) = \lim_n \prod_k (1 - p_{nk} + e^{iz} p_{nk}) \\ &= \lim_n \prod_k (1 + p_{nk}(e^{iz} - 1)). \end{aligned}$$

The last term above equals $\exp\{\lambda(e^{iz} - 1)\}$, since $\sum_k p_{nk} \to \lambda$ and $\sum_k p_{nk}^2 \leq \max_k p_{nk} \cdot \sum_k p_{nk} \to 0$ hold. This implies that the distribution of y is a Poisson distribution.

(ii) The case when some subsequence of $\sum_k p_{nk}$, $n = 1, 2, \cdots$, converges to a finite limit.

We can get the same conclusion as for case (i) above.

(iii) The case when $\sum_k p_{nk} \to \infty$.

As we saw above p_{nk} tends to 0 uniformly in k as $n \to \infty$, and therefore, for an arbitrary $\lambda > 0$ and for each n, we can find $K = K(n, \lambda)$ so that

$$\sum_{k=1}^{K} p_{nk} \to \lambda \ (\text{as } n \to \infty).$$

If we set $y''_n = \sum_{k=1}^{K} y'_{nk}$, then we can conclude as in case (i) that

$$|E(e^{izy})| \leq \lim_n |\prod_{k=1}^{K} E(e^{izy'_{nk}})| = |\exp(\{\lambda(e^{iz} - 1)\})| = \exp\{\lambda(\cos z - 1)\}$$

holds. As λ can be taken as large as we please in the above, we see that the last term above tends to 0 if $0 < z < 2\pi$ when we let $\lambda \to \infty$. Thus we must have $|E(e^{izy})| = 0$. On the other hand by letting $z \downarrow 0$, we see $|E(e^{izy})| \to 1$, which yields a contradiction $1 = 0$. This implies that case (iii) does not occur, and the proof of the theorem is complete. □

2.10. Separable Wiener Processes

The definition of a Wiener process was given and its existence was shown in §2.2. We suppose that the definition and the existence of a separable Wiener process are obvious, and need not be given here.

THEOREM 2.10.1. *The probability that a sample process of a separable Wiener process is continuous is equal to* 1.

PROOF. Let x_t, $t \in T = [a, b)$, be a Wiener process, and let S be the countable subset of T, which appears in the definition of the separability. Namely, the set

$$\Omega' = \{\omega / \varliminf_{\substack{s \to t \\ s \in S}} x_s(\omega) \leq x_t(\omega) \leq \varlimsup_{\substack{s \to t \\ s \in S}} x_s(\omega), \quad t \in T\}$$

has probability 1. From the properties of a normal distribution it follows that

$$P\{\omega / |x_\beta - x_\alpha| > \epsilon\} = o(\beta - \alpha).$$

For $\alpha = \alpha_0 < \alpha_1 < \cdots < \alpha_n = \beta$, $y_\nu = x_{\alpha_\nu} - x_{\alpha_{\nu-1}}$, $\nu = 1, 2, \cdots, n$, are independent, and therefore, by Ottaviani's Theorem, we get

$$P\{\omega / \max_\nu |x_{\alpha_\nu} - x_\alpha| > 2\epsilon\} \leq 2P\{\omega / |x_\beta - x_\alpha| > \epsilon\}.$$

When we have infinitely many points $t_1, t_2, \cdots$ in the interval $[\alpha, \beta]$, we first fix n and rearrange $t_1, t_2, \cdots, t_n$ in an increasing order and apply the result obtained above. Then letting $n \to \infty$, we end up with

$$P\{\omega / \sup_\nu |x_{t_\nu} - x_\alpha| > 2\epsilon\} = o(\beta - \alpha),$$

from which it follows, because of the separability assumption, that

$$P\{\omega / \sup_{\alpha \leq t \leq \beta} |x_t - x_\alpha| > 2\epsilon\} = o(\beta - \alpha).$$

Now if $b < \infty$, then we subdivide the interval $[a, b)$ into n equal sub-intervals $I_{n\nu} = [\alpha_{n,\nu-1}, \alpha_{n,\nu}]$, $\nu = 1, 2, \cdots, n$, and obtain

$$P\{\omega / \max_{\nu=1}^{n} \sup_{t \in I_{n\nu}} |x_t - x_{\alpha_{n,\nu-1}}| > 2\epsilon\}$$
$$\leq \sum_{\nu} P\{\omega / \sup_{t \in I_{n\nu}} |x_t - x_{\alpha_{n,\nu-1}}| > 2\epsilon\} = n \cdot o\left(\frac{1}{n}\right) = o(1).$$

If $|t - s| < (b - a)/n$, then t and s lie either in the same $I_{n\nu}$ or in adjacent ones, and therefore,

$$P\{\omega / \sup_{|t-s|<(b-a)/n} |x_t - x_s| > 4\epsilon\} = o(1).$$

As the set in $\{\ \ \}$ above decreases as $n \to \infty$, we get

$$P\{\omega / \lim_{n} \sup_{|t-s|<(b-a)/n} |x_t - x_s| > 4\epsilon\} = 0.$$

Finally, by letting $\epsilon \downarrow 0$, we obtain the conclusion of the theorem.

In case $b = \infty$, we let

$$\Omega'_k = \{\omega / x_t(\omega) \text{ is continuous in } a \leq t \leq k\},$$
$$\Omega' = \{\omega / x_t(\omega) \text{ is continuous in } a \leq t < \infty\}.$$

Then $\Omega'_k \downarrow \Omega'$ as $k \to \infty$, and from what we obtained above $P(\Omega'_k) = 1$ for each k. Thus we can conclude that $P(\Omega') = 1$ holds as well. □

As was the case for a Poisson process, we can define a Wiener process in a wide sense as an additive process which is continuous in probability and for which the distribution of $x_t - x_s$ is a normal distribution. If we denote by $N(\bullet; m(t), v(t))$ the distribution of x_t, i.e., $x_t - x_a$, then the distribution of $x_t - x_s$ becomes $N(\bullet; m(t) - m(s), v(t) - v(s))$. From the assumption of the continuity in probability, it follows that both $m(t)$ and $v(t)$ are continuous functions of t. It is clear that $v(t)$ is an increasing function. The existence of a Wiener process in a wide sense and that of a separable Wiener process in a wide sense should be obvious, and needs no further explanation.

THEOREM 2.10.2. *The assertion of Theorem* 2.10.1 *is valid also for a separable Wiener process in a wide sense.*

THEOREM 2.10.3. *If, for an additive process which is continuous in probability, its sample process is continuous with probability* 1, *then the process is a separable Wiener process in a wide sense.*

PROOF. Let x_t, $t \in T = [a, b)$, be the additive process satisfying the assumptions of the theorem. The separability of the process is clear from the hypothesis of the continuity with probability 1 of the sample process. Therefore, it suffices to show that the distribution of $x_t - x_s$ $(t > s)$ is a normal distribution. In view of the fact that a continuous function is uniformly continuous over a finite interval, we see that for an arbitrary ϵ there exists a $\delta = \delta(\epsilon)$ such that

$$P\{\omega / u, v \in [t, s], |u - v| < \delta \Rightarrow |x_u - x_v| < \epsilon\} > 1 - \epsilon.$$

Let $\epsilon_1 > \epsilon_2 > \cdots \to 0$, and choose an increasing sequence of natural numbers $p(n)$ in such a way that $(t - s)/p(n) < \delta(\epsilon_n)$. Next, divide the interval $[s, t)$ into

$p(n)$ equal sub-intervals by taking $s = s_{n0} < s_{n1} < \cdots < s_{np(n)} = t$, and let $y_{n\nu} = x_{s_{n\nu}} - x_{s_{n,\nu-1}}$. Define

$$y'_{n\nu} = \begin{cases} y_{n\nu}, & \text{if } |y_{n\nu}| < \epsilon_n, \\ 0, & \text{if } |y_{n\nu}| \geq \epsilon_n, \end{cases}$$

and $y'_n = \sum_\nu y'_{n\nu}$. Then, we see that

$$P\{\omega / x_t - x_s \neq y'_n\} < \epsilon_n,$$

and since $\{y'_\nu\}$, $\nu = 1, 2, \cdots, p(n)$, are independent, we get for $y = x_t - x_s$,

$$E(e^{izy}) = \lim_n E(e^{izy'_n}) = \lim_n \prod_\nu E(^{izy'_{n\nu}}). \tag{2.10.1}$$

Let us denote $m_{n\nu} = E(y'_{n\nu})$, $v_{n\nu} = V(y'_{n\nu})$, $m_n = \sum_\nu m_{n\nu}$, $v_n = \sum_\nu v_{n\nu}$, and consider several possibilities:

(i) The case when $m_n \to m$ (m is finite) and $v_n \to v$ ($< \infty$). In this case we obtain using (2.10.1) that

$$\begin{aligned} E(e^{izy}) &= \lim_n e^{izm_n} \prod_\nu E(e^{iz(y'_{n\nu} - m_{n\nu})}) \\ &= e^{izm} \lim_n \prod_\nu \left(1 - \frac{v_{n\nu}}{2} z^2 + v_{n\nu} \cdot O(\epsilon_n)\right) \\ &= e^{izm} \lim_n \prod_\nu \exp(-\frac{v_{n\nu}}{2} z^2 + v_{n\nu} \cdot O(\epsilon_n)) \\ &= e^{izm} \lim_n \exp(-\frac{v_n}{2} z^2 + v_n \cdot O(\epsilon_n)) \\ &= e^{izm - \frac{v}{2} z^2}. \end{aligned}$$

Therefore, we conclude that the distribution of $y = x_t - x_s$ is a normal distribution.

(ii) The case when some subsequence of m_n and of v_n converges to a finite limit, respectively. We can argue as in case (i) above to get the same conclusion.

(iii) The case when some subsequence of v_n converges to a finite limit v. In this case, we can argue as in case (i) to conclude that the distribution of $y'_n - m_n$ or its subsequence converges to the normal distribution $N(\bullet; 0, v)$. Since the distribution of y'_n converges to that of $y = x_t - x_s$, m_n or its subsequence converges to some finite value m, and hence this case can be reduced to case (ii).

(iv) The case when $v_n \to \infty$. Since $|v_{n\nu}| \leq 4\epsilon^2$, for an arbitrary $v > 0$, we can choose an increasing sequence $\{q(n)\}$ of natural numbers so that $\sum_{\nu=1}^{q(n)} v_{n\nu} \to v$. From (2.10.1) we get

$$|E(e^{izy})| = \overline{\lim_n} \prod_{\nu=1}^{q(n)} |E(e^{izy'_{n\nu}})| = e^{-\frac{v}{2} z^2}.$$

By letting $v \to \infty$, we get $|E(e^{izy})| = 0$, which is a contradiction. Thus we conclude that case (iv) does not occur. □

2.11. Additive Processes Continuous in Probability and Infinitely Divisible Distributions

We have already seen in §2.7 that a general additive process x_t, $t \in T$, can be decomposed into the form

$$x_t = u_t + v_t + g(t),$$

where $\delta(u_t)$ is continuous, $g(t)$ is a function of t alone, $E(\arctan u_t) = 0$, $\delta(v_t)$ is purely discontinuous, $E(\arctan v_t) = 0$, and furthermore, u_t, $t \in T$, and v_t, $t \in T$, are independent. We explained the structure of v_t in §2.7, so it remains to investigate the structure of u_t. As we already explained,

$$P\{\omega/u_{t-} = u_t = u_{t+}\} = 1, \quad t \in T. \tag{2.11.1}$$

Hence it is clear that u_t is continuous in probability. However, as we remarked for a Poisson process, (2.11.1) does not imply that a sample process of u_t is continuous with probability 1. Let us see what an additive process continuous in probability is like.

THEOREM 2.11.1. *Let x_t be an additive process continuous in probability. If, for an arbitrary t, we take an increasing sequence $t_1 < t_2 < \cdots \to t$, then the sequence x_{t_n} converges to x_t, a.e. The same conclusion holds for a decreasing sequence $t_1 > t_2 > \cdots \to t$ also. We should note, however, that the exceptional ω-set where the convergence does not take place depends, in general, on t and also on the choice of the sequences.*

PROOF. From the hypothesis of the continuity in probability it follows that x_{t_n} converges to x_t in probability. If $t_1 < t_2 < \cdots$, then x_{t_n} can be written as $x_{t_n} = x_{t_1} + (x_{t_2} - x_{t_1}) + \cdots + (x_{t_n} - x_{t_{n-1}})$ and the terms on the right-hand side are independent. Therefore, Theorem 2.5.3 in §2.5 implies that x_{t_n} converges a.e., and thus x_{t_n} converges to x_t a.e. The case for $t_1 > t_2 > \cdots$ can be shown in the same way. □

We have seen that a Poisson process in a wide sense and a Wiener process in a wide sense are both continuous in probability, and the distributions of $x_t - x_s$ $(t > s)$ are Poisson distribution and normal distribution, respectively. These facts lead us to conjecture that a general class of distributions, including Poisson and normal distributions, may be obtained by considering the distribution of increments $x_t - x_s$ of a general additive process continuous in probability. Such distributions are called infinitely divisible distributions.

Let us say that a 1-dimensional distribution Φ belongs to the class $U(\epsilon)$ if it satisfies

$$\int_{|\xi|>\epsilon} \Phi(d\xi) < \epsilon.$$

It is easy to see that the fact that a sequence of distributions Φ_n converges to the unitary distribution is equivalent to the statement that for every $\epsilon > 0$ there exists $n_0(\epsilon)$ such that $\Phi_n \in U(\epsilon)$ whenever $n > n_0(\epsilon)$.

Let us denote by Φ_{st} the distribution of the increment $x_t - x_s$ of an additive process continuous in probability. From the continuity in probability it follows that for arbitrary $\epsilon > 0$ and $u \in T$, there exists a $\delta = \delta(\epsilon, u)$ such that

$$|v - u| < \delta \Rightarrow P\{\omega/|x_v - x_u| > \epsilon\} < \epsilon.$$

Because of the Borel Covering Theorem, we see that $\delta(\epsilon, u)$ can be chosen depending only on ϵ and independent of u (we then write simply $\delta(\epsilon)$) as long as $s \leq u \leq t$. If we divide the interval $[s, t)$ into sufficiently small sub-intervals $s = u_0 < u_1 < \cdots < u_n = t$, where $u_i - u_{i-1} < \delta(\epsilon)$, $i = 1, 2, \cdots, n$, then we see from the above that

$$\Phi_{u_{i-1}u_i} \in U(\epsilon), \quad i = 1, 2, \cdots, n,$$

holds. Furthermore, from the additive property of the process we get

$$\Phi_{st} = \Phi_{u_0u_1} * \Phi_{u_1u_2} * \cdots * \Phi_{u_{n-1}u_n}.$$

Roughly speaking this says that Φ_{st} can be represented as a convolution product with factors being arbitrarily close to the unit distribution.

Leaving, for a moment, additive processes, let us call a distribution Φ **infinitely divisible** if for every $\epsilon > 0$, Φ can be decomposed as

$$\Phi = \Phi_1 * \Phi_2 * \cdots * \Phi_n, \quad \Phi_1, \Phi_2, \cdots, \Phi_n \in U(\epsilon).$$

Since

$$N(\bullet; m, v) = N_1 * \cdots * N_n, \quad N_1 = N_2 = \cdots = N_n = N(\bullet; \frac{m}{n}, \frac{v}{n}),$$

$$P(\bullet; \lambda) = P_1 * \cdots * P_n, \qquad P_1 = P_2 = \cdots = P_n = P(\bullet; \frac{\lambda}{n}),$$

both the normal distribution and the Poisson distribution are infinitely divisible. The distribution Φ_{st} of the increment of an additive process continuous in probability, which we discussed above, is clearly infinitely divisible also. Conversely,

THEOREM 2.11.2. *For any infinitely divisible distribution Φ, we can construct an additive process x_t, $t \in [0, 1]$, in such a way that the distribution of $x_1 (= x_1 - x_0)$ equals Φ.*

PROOF. If $\Phi = \delta(\bullet; m)$, then we can set $x_t(\omega) = m \cdot t$; so let us put this case aside, and assume that $\delta(\Phi) > 0$. Let us also call the constant c the **central value** $\alpha(\Phi)$ of Φ, which is uniquely determined by the condition

$$\int \arctan(\xi - c)\Phi(d\xi) = 0.$$

We considered this condition when we discussed Doob's stabilizing sequence in §2.6. If $\alpha(\Phi) = c$, then the central value of $\Phi * \delta(\bullet; m)$ is equal to $c + m$. In proving the theorem, we may assume that $\alpha(\Phi) = 0$ without loss of generality. This is because if Φ is an arbitrary infinitely divisible distribution, then $\Phi * \delta(\bullet; -c)$, where $c = \alpha(\Phi)$, is also infinitely divisible with its central value equal to 0, and, if x_t is an additive process corresponding to the latter, then $y_t = x_t + ct$ would be an additive process corresponding to Φ.

Let S be the set of all rational numbers in $[0, 1]$, and let us define a family $\mathcal{P} = \{\Phi_{st}, s \leq t, s, t \in S\}$ of distributions satisfying the following four conditions:

(i) $\Phi_{01} = \Phi$,
(ii) $s \leq t \leq u \Rightarrow \Phi_{st} * \Phi_{tu} = \Phi_{su}$,
(iii) $\alpha(\Phi_{01}) = 0$,
(iv) $\delta(\Phi_{0t}) = t\delta(\Phi)$.

Let $\{\epsilon_n\}$ be a sequence of positive numbers such that $\epsilon_n \to 0$, and let the decomposition of Φ corresponding to ϵ_n be given by

$$\Phi = \Phi_{n1} * \Phi_{n2} * \cdots * \Phi_{np(n)}, \quad \Phi_{ni} \in U(\epsilon_n).$$

By replacing Φ_{ni} by $\Phi_{ni} * \delta(\bullet; m_i)$ if necessary, we may assume that

$$\alpha(\Psi_{ni}) = 0, \quad \Psi_{ni} = \Phi_{n1} * \cdots * \Phi_{ni}, \quad i = 1, 2, \cdots, n.$$

We may assume also that δ-distributions do not appear among Φ_{ni}. Then $\delta(\Psi_{ni})$ becomes strictly increasing in i, and for $s \in S$, we can determine i so that

$$\delta(\Psi_{n,i-1})/\delta(\Phi) \leq s < \delta(\Psi_{n,i})/\delta(\Phi),$$

and for $t \in S$ $(t > s)$, we can likewise determine j. Then we can set

$$\Phi_{st}^{(n)} = \begin{cases} \Phi_{n,i} * \cdots * \Phi_{n,j-1}, & \text{if } i < j, \\ \text{unit distribution}, & \text{if } i = j. \end{cases}$$

$\mathcal{P}^{(n)} \equiv \{\Phi_{st}^{(n)},\ s \leq t,\ s, t \in S\}$ can be regarded as an approximating family for $\mathcal{P}$. Properties (i), (ii), and (iii) above are satisfied by $\mathcal{P}^{(n)}$, but property (iv) is satisfied only approximately. We intend to choose a suitable subsequence of $\mathcal{P}^{(n)}$, and obtain $\mathcal{P}$ as the limit of this subsequence. In order to carry out such an argument, we need the following lemma.

LEMMA 2.11.1. *If $\Phi_n * \Psi_n = \Phi$, $n = 1, 2, \cdots$, holds, then there exist a subsequence $\{\Phi_{n_p}\}$ of $\{\Phi_n\}$ and a sequence $\{m(n_p)\}$ of real numbers such that $\{\Phi_{n_p} * \delta(\bullet; m(n_p))\}$ converges to some distribution Φ_∞. If, in particular, $\alpha(\Phi_n) = 0$, we can take $m(n_p) = 0$.*

PROOF. Let $F_n(x)$ be the distribution function of Φ_n. Since it follows from the assumption $\Phi_n * \Psi_n = \Phi$ that $[\Phi_n * \delta(\bullet; m)] * [\Psi_n * \delta(\bullet; -m)] = \Phi$ holds, we may assume without loss of generality that

$$F_n(-0) \leq 1/2, \quad F_n(0) \geq 1/2$$

are satisfied. From this it follows that

$$F_n(y) - F_n(x) > 1/2 \Rightarrow x \leq 0 \leq y. \tag{2.11.2}$$

From the assumption $\Phi_n * \Psi_n = \Phi$ we get

$$\int_{-\infty}^{\infty} [F_n(\ell - x) - F_n(-\ell - x)]\Psi_n(dx) = \Phi[-\ell, \ell].$$

If we take ℓ sufficiently large so that the right-hand side of the equation above becomes greater than $1/2$, then for a suitable real number $m(n)$, we get

$$F_n(\ell - m(n)) - F_n(-\ell - m(n) - 0) \geq 1/2.$$

Then, in view of (2.11.2), we get $-\ell \leq m(n) \leq \ell$ for all n. Therefore, we can choose a suitable subsequence $\{n_p\}$, $p = 1, 2, \cdots$, of $\{n\}$ for which $\{m(n_p)\}$, $p = 1, 2, \cdots$, converges to some m, and then choose a further subsequence of $\{n_p\}$, which we denote again by $\{n_p\}$, in such a way that $\{F_{n_p}(x - m(n_p))\}$ converges to a monotone increasing right-continuous function $F_0(x)$ at every x except for an at most countable number of exceptional points. Since $m(n_p) \to m$ holds also, we see that $\{F_{n_p}(x - m)\}$ converges, as $p \to \infty$, to $F_0(x)$ in the same way as above. Thus, we can conclude that $\{F_{n_p}\}$ converges to $F(x) \equiv F_0(x + m)$ for all x but at most countable exceptional points. We also have $0 \leq F(x) \leq 1$, and

$$F(\infty) - F(-\infty) \geq F_0(\ell) - F_0(-\ell - 0) \geq \Phi[-\ell, \ell].$$

Next, let L be an arbitrary number larger than ℓ, and repeat the same arguments as above using L and $\{F_{n_p}(x)\}$. Then, we obtain a further subsequence $\{F_{n'_p}(x)\}$

of $\{F_{n_p}(x)\}$, which converges at every x except for at most countable exceptional points to some function $G(x)$, for which

$$G(\infty) - G(-\infty) \geq \Phi[-L, L].$$

As $\{F_{n_p}(x)\}$ converges to $F(x)$, the subsequence $\{F_{n'_p}(x)\}$ converges to $F(x)$ as well, and therefore, we conclude that $G = F$ holds, and we get

$$F(\infty) - F(-\infty) \geq \Phi[-L, L].$$

As L can be taken arbitrarily large, we can let $L \uparrow \infty$ to obtain that $F(\infty) - F(-\infty) = 1$. This implies that F corresponds to some 1-dimensional distribution Φ_∞, and therefore, the sequence $\{\Phi_{n_p}\}$ converges to Φ_∞. In case $\alpha(\Phi_n) = 0$, we get from $\Phi_{n_p} * \delta(\bullet; m(n_p)) \to \Phi_\infty$ that $m(n_p) \to \alpha(\Phi_\infty)$, which implies that $\{\Phi_{n_p}\}$ itself converges to Φ_∞. □

Let us continue with the proof of the theorem. Since $\Phi_{st}^{(n)} * [\Phi_{0t}^{(n)} * \Phi_{t1}^{(n)}] = \Phi$ holds, by Lemma 2.11.1 there exist a subsequence $\{n_p\}$ of $\{n\}$ and a sequence $\{m_{n_p}(s,t)\}$ of real numbers such that $\{\Phi_{st}^{(n_p)} * \delta(\bullet; m_{n_p}(s,t))\}$ converges as $p \to \infty$. The choice of the subsequence $\{n_p\}$ shown in the proof of Lemma 2.11.1 depended on s, t $(\in S)$, but using the fact that S is a countable set, we can use the diagonal selection argument to choose a single subsequence $\{n_p\}$, which works for all the pairs s, t $(\in S)$. From the remark given in the last part of Lemma 2.11.1, we see that we can take $m_{n_p}(0,t)$ to be 0. We shall show that $m_{n_p}(s,t)$ can be taken to be 0. For this it is enough to show that $m_{n_p}(s,t)$ converges as $p \to \infty$. Clearly, we have

$$\Phi_{0s}^{(n_p)} * [\Phi_{st}^{(n_p)} * \delta(\bullet; m_{n_p}(s,t))] = \Phi_{0t}^{(n_p)} * \delta(\bullet; m_{n_p}(s,t)).$$

By the choice of the subsequence $\{n_p\}$, the left-hand side above and $\Phi_{0t}^{(n_p)}$ on the right-hand side converge as $p \to \infty$. Therefore, $\{m_{n_p}(s,t)\}$ must also converge. Consequently, we can conclude that $\{\Phi_{st}^{(n_p)}\}$ itself converges. Let us denote this limit Φ_{st}. Then we get $\Phi_{st} * \Phi_{tu} = \Phi_{su}$, when $s \leq t \leq u$, since $\Phi_{st}^{(n)} * \Phi_{tu}^{(n)} = \Phi_{su}^{(n)}$ holds for each n. As $\alpha(\Phi_{0t}^{(n)}) = 0$, $\alpha(\Phi_{0t}) = 0$ holds also. From

$$\delta(\Phi_{0t}^{(n)}) \leq t \cdot \delta(\Phi) < \delta(\Phi_{0t}^{(n)} * \Phi_{n,i(n,t)}), \quad \text{with } \Phi_{n,i(n,t)} \in U(\epsilon_n),$$

where $\{\Phi_{n,i(n,t)}\}$ tends to the unit distribution, we see, as $n \to \infty$, that we have

$$\delta(\Phi_{0t}) \leq t \cdot \delta(\Phi) \leq \delta(\Phi_{0t}) \quad \text{and hence } \delta(\Phi_{0t}) = t \cdot \delta(\Phi).$$

Thus we can conclude that $\mathcal{P} = \{\Phi_{st}; s < t,\ s, t \in S\}$ satisfies properties (i), (ii), (iii), and (iv) of the theorem.

By Theorem 2.1.2 in §2.1 there exists an additive process y_t, $t \in S$, such that the distribution of $y_t - y_s$ $(t > s)$ is Φ_{st}. (Although the proof given for Theorem 2.1.2 treated the case when t varies in T, which is some interval, the same argument works for the case $T = S$.) If for an arbitrary $t \in T = [0,1]$ we take a sequence of points $t_n \uparrow t$ from S, then $\{y_{t_n}\}$ becomes an additive sequence, for which $\alpha(y_{t_n}) = 0, \delta(y_{t_n}) = t_n \cdot \delta(\Phi) \leq t \cdot \delta(\Phi)$ hold. Therefore, $\{y_{t_n}\}$ converges a.e. If we denote this limit by x_t, then $x_t, 0 \leq t \leq 1$, becomes an additive process, for which $\alpha(x_t) = 0, \delta(x_t) = t \cdot \delta(\Phi)$ are satisfied. This implies that $\delta(x_t)$ is continuous in t, and hence, x_t is an additive process continuous in probability. □

2.12. Structure of Separable Additive Processes Continuous in Probability

As we pointed out in §2.8, it suffices, when we deal with additive processes, to discuss separable additive processes. It is the purpose of this section to investigate in detail the structure of separable additive processes continuous in probability. Since most of the proofs of the assertions made below are quite complicated technically, we shall omit them.

THEOREM 2.12.1. *If x_t, $t \in [a, b)$, is a separable additive process continuous in probability, then with probability* 1, *its sample process is a discontinuous function of the first kind.*

A function $f(t)$ is called a discontinuous function of the first kind if at every t, both $f(t-0)$ and $f(t+0)$ exist. Suppose we draw the graph of

$$u = f(t+0) - f(t-0), \quad a \leq t \leq b,$$

in the (t, u)-plane. What we mean by a "graph" here refers, in general, to a discontinuous set of points, which is a subset $G(f)$ of the strip $D : a \leq t \leq b,\ -\infty < u < \infty$. Let us denote by $N_f(E)$ the number of points contained in $E \cap G(f)$ for an arbitrary Borel subset E of D. If $E \cap G(f)$ is an infinite set, then we let $N_f(E) = \infty$ regardless of the cardinality of this set. $N_f(E)$ counts the number of points of the graph contained in E and can be considered to be a measure defined on D. If the set E is **strictly separated** from the t-axis (i.e., has positive distance away from the t-axis), then, using the hypothesis that $f(t)$ is a discontinuous function of the first kind, it can be proved easily that $N_f(E)$ is finite. Next, let us denote by $S_f(E)$ the sum of the u-coordinates of all the points in the set $E \cap G(f)$:

$$\begin{aligned} S_f(E) &= \sum_{(t,u) \in E \cap G(f)} u = \sum_{(t, f(t+0)-f(t-0)) \in E} (f(t+0) - f(t-0)) \\ &= \int_E u N_f(dt\, du). \end{aligned}$$

If E is contained entirely in the upper-half plane or lower-half plane, then the value of $S_f(E)$ is determined (it may take either $+\infty$ or $-\infty$), but if E contains a part of the t-axis, then the value of $S_f(E)$ may not be defined. But if E is strictly separated from the t-axis, then $S_f(E)$ takes a definite finite value.

When $x_t(\omega)$, $t \in [a, b)$, is a separable additive process continuous in probability, then, with probability 1, its sample process is a discontinuous function of the first kind, so $N_x(E)$ and $S_x(E)$ discussed above can be defined and depend on ω; in fact, it is a measurable function of ω.

THEOREM 2.12.2. *The distribution of $N_x(E)$ is a Poisson distribution. Here we assume that the distribution of a random variable always taking the value ∞ is given by $P(\bullet; \infty)$.*

THEOREM 2.12.3. *When the sets $E_1, E_2, \cdots$ are pairwise disjoint, then $N_x(E_1)$, $N_x(E_2), \cdots$ are independent random variables, and if we let $E = \bigcup_n E_n$, then*

$$N_x(E) = \sum_n N_x(E_n).$$

Let us denote by $\mathcal{N}$ the family of random variables $\{N_x(E), E \in \boldsymbol{B}(D)\}$. From the two theorems above we can see that $\mathcal{N}$ resembles a Poisson process. Accordingly, we call $\mathcal{N}$ a **Poisson additive system**. If we denote by $n(E)$ the mean value of $N_x(E)$, then the distribution of the random variable $N_x(E)$ is given by $P(\bullet; n(E))$. From the additivity of $N_x(E)$ it follows that $n(E)$ is a measure on the strip $D : a \leq t \leq b,\ -\infty < u < \infty$. If E is strictly separated from the t-axis, $N_x(E)$ takes only finite values, and therefore, $n(E)$ is also finite. If E is not separated from the t-axis, then $n(E)$ may take the value ∞, but its magnitude is controlled by the following theorem.

THEOREM 2.12.4.

$$\int_{u=-1}^{1} \int_{t=a}^{b} u^2 n(dt\, du) < \infty.$$

Let $S_n(t)$ be the total of the magnitudes of jumps of x_t bigger than $1/n$ in absolute value summed over a to t, namely,

$$S_n(t) = \int_{|u|>1/n} \int_{\tau=a}^{t} uN(d\tau\, du).$$

Then $S_n(t)$ is an additive process, and even though $S_n(t)$ does not converge in general when $n \to \infty$, we can make it converge by subtracting a suitable function of t, which does not depend on ω. Namely, we have the following:

THEOREM 2.12.5. *If we set*

$$S_n^*(t) = S_n(t) - \int_{|u|>1/n} \int_{\tau=a}^{t} \frac{u}{1+u^2} n(d\tau\, du),$$

then, with probability 1, $S_n^*(t)$ *converges uniformly in* t *as* $n \to \infty$. *If we denote this limit by* $S(t)$, *then* $S(t)$ *is also an additive process, and the graph of* $u = S(t+0) - S(t-0)$ *coincides with the graph of* $x_{t+0} - x_{t-0}$ *with probability* 1.

From this theorem it follows that, with probability 1, $y_t = x_{t+0} - S(t+0)$ is a continuous function of t. Furthermore,

THEOREM 2.12.6. y_t *is a separable Wiener process, and* y_t *and* $\mathcal{N} = \{N_x(E), E \in \boldsymbol{B}(D)\}$ *mentioned above are independent. Summarizing the results explained so far, we have*

$$x_{t+0} = y_t + \lim_n \int_{|u|>1/n} \int_{\tau=a}^{t} \left[uN(d\tau\, du) - \frac{u}{1+u^2} n(d\tau\, du) \right].$$

2.13. Canonical Form of Infinitely Divisible Distributions

Let Φ be an infinitely divisible distribution. As we have seen in §2.11, Φ can be regarded as the distribution of x_1 of some additive process x_t, $0 \leq t \leq 1$, which is continuous in probability. We may assume that x_t is separable, by taking its separable modification if necessary. From the result explained in the preceding section, we know that

$$x_{t+0} = y_t + \lim_n \int_{|u|>1/n} \int_{\tau=a}^{t} \left[uN(d\tau\, du) - \frac{u}{1+u^2} n(d\tau\, du) \right].$$

Setting $m(t) = E(y_t), v(t) = V(y_t)$, we obtain, by using the fact that $\mathcal{N}$ is a Poisson additive system and that $\{y_t\}$ is a separable Wiener process independent of $\mathcal{N}$, that

$$E(e^{izx_{t+0}}) = \exp\left\{ im(t) - \frac{v(t)}{2}z^2 + \lim_{n\to\infty}\int_{|u|>1/n}\int_{\tau=a}^{t}\left(e^{izu} - 1 - \frac{izu}{1+u^2}\right) n(d\tau\, du)\right\},$$

which can be rewritten, by setting $n_t(du) = \int_{\tau=a}^{t} n(d\tau\, du)$, as

$$E(e^{izx_{t+0}}) = \exp\left\{ im(t) - \frac{v(t)}{2}z^2 + \lim_{n\to\infty}\int_{|u|>1/n}\left(e^{izu} - 1 - \frac{izu}{1+u^2}\right) n_t(du)\right\}.$$

If we set $t = 1$, then the right-hand side of the equation above becomes $E(e^{izx_1})$, which is the characteristic function $\varphi(z)$ of the distribution Φ of x_1. By writing $m(1) = m, v(1) = v$, and $n_1(E - \{0\}) = n(E)$, we get the following:

$$\varphi(z) = \exp\left\{ imz - \frac{v}{2}z^2 + \int_{-\infty}^{\infty}\left(e^{izu} - 1 - \frac{izu}{1+u^2}\right) n(du)\right\}. \tag{2.13.1}$$

Here, m is a real number, $v \geq 0$, and n is a measure on R^1 such that

$$n(\{0\}) = 0, \quad \int_{-\infty}^{\infty}\frac{u^2}{1+u^2}n(du) < \infty.$$

This is the canonical form due to P. Lévy for the characteristic function of an infinitely divisible distribution. Conversely, when $m, v, n(du)$ satisfying the conditions stated above are given, define a function $\varphi(z)$ by using (2.13.1). Then $\varphi(z)$ becomes the characteristic function of some infinitely divisible distribution. In order to show that a function defined by (2.13.1) is a characteristic function, it is enough to note the following. Let us denote by $\boldsymbol{\Psi}$ the set of all functions $\psi(z)$ for which $\exp(\psi(z))$ is a characteristic function. Then

(i) From the form of characteristic functions of delta, normal, and Poisson distributions, we can see that $imz, -\frac{v}{2}z^2, \lambda(e^{iz} - 1) \in \boldsymbol{\Psi}$,

(ii) $\psi(z) \in \boldsymbol{\Psi} \Rightarrow \psi(zu) \in \boldsymbol{\Psi}$,

(iii) $\psi_1(z), \psi_2(z), \cdots, \psi_n(z) \in \boldsymbol{\Psi} \Rightarrow \sum_i \psi_i(z) \in \boldsymbol{\Psi}$,

(iv) $\psi_n(z) \in \boldsymbol{\Psi}, \psi_n(z) \to \psi(z)$ (uniformly in every compact set) $\Rightarrow \psi(z) \in \boldsymbol{\Psi}$.

In order to see that the characteristic function $\varphi(z)$ defined by (2.13.1) corresponds to an infinitely divisible distribution, let us denote by $\varphi_n(z)$ the function defined by (2.13.1) by using $m/n, v/n, n(du)/n$ in place of $m, v, n(du)$, respectively. Then, we see that $\varphi(z) = \varphi_n(z)^n$, and furthermore, as $n \to \infty$, we have that $\varphi_n(z) \to 1$ (uniformly over every compact set), and thus, if we denote by Φ, Φ_n the distributions corresponding to $\varphi(z), \varphi_n(z)$, respectively, then

$$\Phi = \Phi_n * \Phi_n * \cdots * \Phi_n (n \text{ terms}), \quad \Phi_n \to \text{unit distribution}$$

holds as $n \to \infty$, which shows that Φ is infinitely divisible.

We have shown that a necessary and sufficient condition for a function $\varphi(z)$ to be the characteristic function of some infinitely divisible distribution is that $\varphi(z)$ can be represented in the form (2.13.1). We can show furthermore that the quantities $m, v, n(du)$ appearing in (2.13.1) are determined uniquely by the given distribution. However, we will omit the proof of this fact here.

Normal distributions correspond with the case where $n(du) \equiv 0$, and Poisson distributions with the case where $v = 0$ and $n(du)$ has the support consisting of a single point $\{1\}$ with $n(\{1\}) = \lambda, m = \lambda/2$.

We already mentioned the fact that probability laws given by infinitely divisible distributions appear as distributions for increments of some additive processes. We can in fact show that they appear as distributions for increments of temporally homogeneous additive processes. This fact can be seen in the following way. If we denote the characteristic function of an infinitely divisible distribution Φ in the form of (2.13.1) and if we represent it as $\varphi(z) = \exp(\psi(z))$, then $\varphi_t(z) = \exp\{t\psi(z)\}$ also gives a characteristic function. If we denote by Φ_{st} $(0 \leq s \leq t \leq 1)$ the distribution function corresponding to $\varphi_{t-s}(z)$, then since $\varphi_s(z)\varphi_t(z) = \varphi_{s+t}(z)$ holds, we get $\Phi_{st} * \Phi_{tu} = \Phi_{su}$ $(s \leq t \leq u)$. Therefore, there exists an additive process x_t $(0 \leq t \leq 1)$ for which the distribution of $x_t - x_s$ $(s \leq t)$ is given by Φ_{st}. As the distribution Φ_{st} depends only on the difference $t - s$, the process x_t is temporally homogeneous.

(2.13.1) can be represented also in the following form:

$$\varphi(z) = \exp\left\{imz + \int_{-\infty}^{\infty}\left(e^{izu} - 1 - \frac{izu}{1+u^2}\right)\frac{1+u^2}{u^2}G(du)\right\}, \tag{2.13.2}$$

where $G(du)$ is a finite measure on the real line R^1. The function appearing as the integrand above is not defined when $u = 0$, but it is taken to be the value $-z^2/2$, which is the limit as $u \to 0$ of the integrand. Hence, G-measure $G(\{0\})$ of the single point 0 corresponds to v of (2.13.1). The form (2.13.2) was given by A. Khinchin.

When the variance $V(\Phi)$ of Φ is finite, (2.13.1) can also be written as

$$\varphi(z) = \exp\left\{imz - \frac{v}{2}z^2 + \int_{-\infty}^{\infty}(e^{izu} - 1 - izu)n(du)\right\}, \tag{2.13.3}$$

where $\int u^2 n(du) < \infty$. m appearing in this formula differs from m in (2.13.1), but $v, n(du)$ are the same. This is the form given by Kolmogorov. When

$$\int_{-1}^{1}|u|n(du) < \infty,$$

it can be written also as

$$\varphi(z) = \exp\left\{imz - \frac{v}{2}z^2 + \int_{-\infty}^{\infty}(e^{izu} - 1)n(du)\right\}. \tag{2.13.4}$$

In this case also, m differs from that of (2.13.1), but $v, n(du)$ are the same.

When the corresponding additive process moves only by jumps,

$$\varphi(z) = \exp\left\{\int_{-\infty}^{\infty}(e^{izu} - 1)n(du)\right\}, \tag{2.13.5}$$

and if the process moves only by positive jumps, then

$$\varphi(z) = \exp\left\{\int_{0}^{\infty}(e^{izu} - 1)n(du)\right\}. \tag{2.13.6}$$

2.14. Various Methods for Construction of Poisson Processes

The Poisson process was originally introduced as an integer-valued stochastic process which increases spontaneously and independently over non-overlapping time intervals and satisfies the following properties:

$$\begin{aligned} &P(\Delta x_t = 0) = 1 - \lambda \cdot \Delta t + o(\Delta t), \quad P(\Delta x_t = 1) = \lambda \cdot \Delta t + o(\Delta t), \\ &P(\Delta x_t = k) = o(\Delta t), \quad k \geq 2. \end{aligned} \tag{2.14.1}$$

If the process is not temporally homogeneous, we can replace $\lambda \cdot \Delta t$ by $\Delta\lambda(t)$. For example, the number of accidents a taxi driver causes is considered to be describable by a Poisson process as a first degree approximation. If the amount of traffic is small, then the number of accidents is small, while in a rush-hour the number increases, and therefore, the process counting the number of accidents should not be temporally homogeneous. Since the driver becomes more cautious after causing an accident, the assumption of independent increment may not be quite accurate. If we consider this fact, we may get a better second degree approximation, but we shall not dwell on this point here.

The assumption of independent increments leads to an additive process, and therefore, when it is temporally homogeneous, we get from (2.13.6) that

$$\varphi_{st}(z) = E(e^{iz(x_t - x_s)}) = \exp\left\{(t-s)\int_0^\infty (e^{izu} - 1) n(du)\right\}.$$

Furthermore, since the magnitudes of the jumps are positive integers, the support of $n(du)$ is the set $\{1, 2, 3, \cdots\}$, and we have

$$\begin{aligned} \varphi_{st}(z) &= \exp\left\{(t-s)\sum_k (e^{ikz} - 1)c_k\right\} \qquad \left(\sum_{k=1}^\infty c_k = \int_1^\infty n(du) < \infty\right) \\ &= 1 - \left(\sum_k c_k\right)(t-s) + (t-s)\sum_k e^{ikz} \cdot c_k + o(t-s). \end{aligned}$$

Therefore, from (2.14.1) it follows that $c_2 = c_3 = \cdots = 0$, and we get

$$\varphi_{st}(z) = \exp((t-s)(e^{iz} - 1)c_1).$$

This means that the distribution of $x_t - x_s$ is a Poisson distribution. The case where the process is not temporally homogeneous can be treated in a similar manner.

Let $x_t(\omega)$, $t \in [0, \infty)$, be a separable, temporally homogeneous Poisson process, and let $T_0(\omega), T_0(\omega) + T_1(\omega), T_0(\omega) + T_1(\omega) + T_3(\omega), \cdots$ be the sequence of times of the increase of the sample process $x_t(\omega)$. $T_0(\omega), T_1(\omega), T_2(\omega), \cdots$ are the sojourn times at $0, 1, 2, \cdots$, respectively. Each of $T_0, T_1, T_2, \cdots$ has the same distribution (called the **exponential distribution**). Namely,

$$P\{\omega / T_n > t\} = e^{-\lambda t}, \quad n = 0, 1, 2, \cdots,$$

and they are independent. For T_0, we have

$$P\{\omega / T_0 > t\} = P\{\omega / x_t = 0\} = e^{-\lambda t}.$$

Thus the density function for the distribution of T_0 is given by $\lambda e^{-\lambda t}dt$. Next, let us derive the distribution of the random vector (T_0, T_1). Let $t_1 > t_0$. Then

$$\begin{aligned} P\{\omega/T_0 > t_0, T_0 + T_1 > t_1\} &= P\{\omega/x_{t_0} = 0, x_{t_1} = 0 \text{ or } 1\} \\ &= P\{\omega/x_{t_0} = 0, x_{t_1} = 0\} + P\{\omega/x_{t_0} = 0, x_{t_1} = 1\} \\ &= e^{-\lambda t_1} + e^{-\lambda t_0} e^{-\lambda(t_1 - t_0)} \cdot \lambda(t_1 - t_0) \\ &= e^{-\lambda t_1}[1 + \lambda(t_1 - t_0)]. \end{aligned}$$

Therefore, for any measurable function f on R^2

$$\begin{aligned} E\{f(T_0, T_1)\} &= E\{f(T_0, T_0 + T_1 - T_0)\} \\ &= \int\int_{t_1 > t_0} f(t_0, t_1 - t_0) \frac{\partial^2 (e^{-\lambda t_1}(1 + \lambda(t_1 - t_0)))}{\partial t_1 \partial t_0} dt_0 dt_1 \\ &= \int\int_{t_1 > t_0} f(t_0, t_1 - t_0) \lambda^2 e^{-\lambda t_1} dt_0 dt_1 \\ &= \int_{t_1=0}^{\infty} \int_{t_0=0}^{\infty} f(t_0, t_1) \lambda^2 e^{-\lambda(t_0 + t_1)} dt_0 dt_1. \end{aligned}$$

This shows that the density function of the distribution of the random vector (T_0, T_1) is given by $\lambda e^{-\lambda t_0} \cdot \lambda e^{-\lambda t_1}$. Therefore, the distributions of T_0 and T_1 are the same exponential distribution, and T_0 and T_1 are independent. We can show the same facts for $T_0, T_1, \cdots, T_n$ similarly.

By utilizing these properties of the sojourn times, we can construct a Poisson process in the following way. Let $T_0, T_1, T_2, \cdots$ be a sequence of independent random variables, and suppose that

$$P\{\omega/T_n > t\} = e^{-\lambda t}, \quad n = 0, 1, 2, \cdots.$$

If we then let

$$x_t = \inf\{n/T_0 + T_1 + \cdots + T_n > t\},$$

x_t becomes a Poisson process. Let us look at a concrete example of a Poisson process obtained from this point of view. Suppose that the lifetime of an electric bulb obeys an exponential distribution. Namely, the probability of the bulb burning out in the time period $[t, t + dt]$ is given by $\lambda e^{-\lambda t}dt$. Suppose we replace a bulb immediately when it burns out, and denote by x_t the number of replacements necessary during the time interval $[0, t]$; then x_t becomes a Poisson process. This method of construction cannot be applied for the case where the process to be obtained is not temporally homogeneous.

Another interesting method of construction of a Poisson process is the following. This is based on a random variable x with a Poisson distribution, and a sequence of independent random variables $y_1, y_2, \cdots$, each having the uniform distribution on the unit interval $[0, 1]$. We assume that x is independent from $y_1, y_2, \cdots$ as well. Denote by $c_t(\xi)$ the indicator function of the interval $[0, t]$, and let

$$x_t = \sum_{i=1}^{x} c_t(y_i);$$

then x_t gives a Poisson process. In fact, let $0 = t_0 < t_1 < \cdots < t_n = 1$, and consider the distribution of the random vector $(x_{t_1} - x_{t_0}, x_{t_2} - x_{t_1}, \cdots, x_{t_n} - x_{t_{n-1}})$. Then we have

$$P\{\omega/x_{t_i} - x_{t_{i-1}} = k_i,\ i = 1, 2, \cdots, n\}$$

$$= P\left[\bigcap_{i=1}^{n}\{\omega/x = \sum k_i,\ \text{exactly } k_i \text{ among } y_1, y_2, \cdots, y_n \text{ are contained in } [t_{i-1}, t_i]\}\right]$$

$$= e^{-\lambda} \cdot \frac{\lambda^k}{k!} \frac{k!}{k_1! \cdots k_n!} \prod_i (t_i - t_{i-1})^{k_i} \quad \left(k = \sum k_i\right)$$

$$= \prod_i e^{-\lambda(t_i - t_{i-1})} \frac{(\lambda(t_i - t_{i-1}))^{k_i}}{k_i!}.$$

This says that x_t, $0 \le t \le 1$, is a Poisson process. We can apply this method of construction even when the process to be obtained is not temporally homogeneous.

2.15. Compound Poisson Processes

Let us consider a temporally homogeneous additive process x_t, $0 \le t < \infty$, which changes only by jumps. Clearly,

$$\varphi_{st}(z) \equiv E(e^{iz(x_t - x_s)}) = \exp\left\{(t - s)\int_{-\infty}^{\infty}(e^{izu} - 1)n(du)\right\}, \tag{2.15.1}$$

where n satisfies

$$\int_{|u|>1} n(du) < \infty, \quad \int_{|u|\le 1} |u| n(du) < \infty. \tag{2.15.2}$$

In this case, the number of jumps, with magnitude whose absolute value exceeds a fixed positive number, occurring within a finite time interval is finite with probability 1, but in general there may be infinitely many jumps of magnitude close to 0. However, since the sum of the magnitudes of the jumps is absolutely convergent with probability 1, we are assured that $x_t - x_s$ is determined as the sum of the magnitudes of jumps. If the following condition which is stronger than (2.15.2) is satisfied,

$$\lambda = \int_{-\infty}^{\infty} n(du) < \infty, \tag{2.15.3}$$

then the number of jumps occurring within a finite time interval is finite with probability 1. An additive process satisfying this condition is called a compound Poisson process.

If we write

$$\Phi(du) = \lambda^{-1} n(du),$$

then Φ becomes a probability distribution on R^1. (2.15.1) can be rewritten in this case as

$$\varphi_{st}(z) = exp\left\{\int_s^t \int_{-\infty}^{\infty} (e^{izu} - 1)\Phi(du) \cdot \lambda d\tau\right\}.$$

From this identity we can see that the probability that there is a jump occurring in the time interval $d\tau$ is given by

$$\int_{u=-\infty}^{\infty} \Phi(du)\lambda \cdot d\tau = \lambda \cdot d\tau,$$

and the probability that the magnitude of this jump lies in du equals $\Phi(du) \cdot \lambda \cdot d\tau$. In this sense we are justified to regard $\Phi(du)$ as representing the distribution of magnitudes of jumps.

If we denote by N_t the number of jumps x_t goes through within the time interval $[0, t]$, then N_t is a Poisson process and

$$\psi_{st}(z) \equiv E(e^{iz(N_t - N_s)}) = \exp\{(t-s)\lambda(e^{iz} - 1)\}.$$

Furthermore, x_t and N_t jump at exactly the same time points. The magnitude of a jump of N_t is always 1, but that of x_t is distributed according to Φ.

Rewriting the expression for $\varphi_{st}(z)$, we get

$$\begin{aligned}\varphi_{st}(z) &= e^{-\lambda(t-s)(1-\phi(z))} \\ &= e^{-\lambda(t-s)} \sum_n \frac{(\lambda(t-s))^n}{n!} \phi(z)^n, \quad \phi(z) = \int e^{izu} \Phi(du).\end{aligned}$$

Therefore, the distribution Φ_{st} of $x_t - x_s$ is given by

$$\begin{aligned}\Phi_{st} &= \sum_n e^{-\lambda(t-s)} \frac{(\lambda(t-s))^n}{n!} \Phi^{*n} \\ &= \sum_n P(n; \lambda(t-s)) \cdot \Phi^{*n}, \quad \Phi^{*n} = \Phi * \cdots * \Phi \ (n \text{ terms}).\end{aligned} \tag{2.15.4}$$

Based on this viewpoint, we can construct a compound Poisson process in the following way. Let $T_1, T_2, \cdots, u_1, u_2, \cdots$ be sequences of independent random variables for which

$$P\{\omega / T_n > t\} = e^{-\lambda t}, \quad P\{\omega / u_n \in du\} = \Phi(du).$$

Define x_t by

$$x_t = u_1 + u_2 + \cdots + u_n \ (T_1 + \cdots + T_n \le t < T_1 + \cdots + T_{n+1}) \ (n = 1, 2, \cdots).$$

Then x_t, $0 \le t < \infty$, becomes a **compound Poisson process** described above. $T_1, T_2, \cdots$ give the **waiting times** for the next jump and $u_1, u_2, \cdots$ describe the magnitudes of the next jump.

As a concrete example of a compound Poisson process, we can mention the accumulated amount of compensations for damages due to accidents committed by a taxi driver. As we explained before, the total number of accidents committed by the driver within the time period $[0, t]$ can be described by a Poisson process (assumed to be temporally homogeneous), and if we let Φ represent the distribution of the amount of compensation for the damage caused by a single accident, then the distribution of accumulated compensations for damages caused by the driver in the time period $[s, t]$ is given by (2.15.4).

2.16. Stable Distributions and Stable Processes

We say that two distributions Φ and Ψ are of the **same type** if there exists a $\lambda > 0$ such that

$$\Psi(E) = \Phi(\lambda E), \quad \lambda E = \{\lambda \cdot \xi; \xi \in E\} \tag{2.16.1}$$

holds for every Borel set $E \subset R^1$. If we denote by $F, G; \varphi, \psi$ the distribution function and the characteristic function of Φ, Ψ, respectively, then condition (2.16.1) is equivalent to

$$G(x) = F(\lambda x) \tag{2.16.2}$$

or

$$\psi(\lambda z) = \varphi(z). \tag{2.16.3}$$

Distribution functions of a random variable x and $\lambda \cdot x$ ($\lambda > 0$) are of the same type. Normal distribution $N(\bullet; 0, v)$ with its mean 0 and $v > 0$ is of the same type as the standard normal distribution $N(\bullet; 0, 1)$. When the convolution $\Phi_1 * \Phi_2$ of an arbitrary pair Φ_1, Φ_2, each of which is of the same type as Φ, is again of the same type as Φ, then Φ is said to be **stable**. If Φ is stable, so is any distribution which is of the same type. The standard normal distribution is stable. We can characterize the stability of a distribution Φ in terms of its characteristic function in the following way: For an arbitrary pair of positive numbers λ_1, λ_2 there exists a $\lambda = \lambda(\lambda_1, \lambda_2) > 0$ such that

$$\varphi(\lambda z) = \varphi(\lambda_1 z)\varphi(\lambda_2 z). \tag{2.16.4}$$

Using this we can show the following:

THEOREM 2.16.1. *A stable distribution is infinitely divisible.*

PROOF. Let $\varphi(z)$ be the characteristic function of a stable distribution Φ. Then from (2.16.4) it follows that there exists $a > 0$ such that $\varphi(az) = \varphi(z)^2$, and hence

$$\varphi(a^n z) = \varphi(z)^{2^n}. \tag{2.16.5}$$

If $a = 1$, then we have $\varphi(z) = \varphi(z)^2$, so that we must have $\varphi(z) = 1$ or $\varphi(z) = 0$. As $\varphi(z)$ is continuous and $\varphi(0) = 1$, we must have $\varphi(z) \equiv 1$. This implies that Φ is the unit distribution and clearly it is infinitely divisible. If $a < 1$, then we can show that $|\varphi(z)| \equiv 1$ must hold. For if there exists a z for which $|\varphi(z)| < 1$, then letting $n \to \infty$ in (2.16.5), we are led to the contradiction $1 = 0$. From $|\varphi(z)| \equiv 1$, we obtain $\delta(\Phi) = 0$ so that Φ is a δ-distribution, which is also infinitely divisible. Finally, if $a > 1$, by substituting z/a^n for z in (2.16.5), we obtain

$$\varphi(z) = \varphi(a^{-n} z)^{2^n}. \tag{2.16.6}$$

We note that $\varphi(a^{-n}z)$ is a characteristic function, and as $n \to \infty$, it converges to the constant 1 uniformly over every compact set, and thus (2.16.6) implies that Φ is an infinitely divisible distribution. □

From this theorem it follows that $\varphi(z)$ can be represented as

$$\varphi(z) = e^{\psi(z)}, \quad \psi(z) = imz - \frac{v}{2}z^2 + \int \left(e^{izu} - 1 - \frac{izu}{1+u^2}\right) n(du), \tag{2.16.7}$$

and the identity (2.16.4) takes the form

$$\psi(\lambda z) = \psi(\lambda_1 z) + \psi(\lambda_2 z). \tag{2.16.8}$$

Using (2.16.8), we obtain further

THEOREM 2.16.2. $\psi(z) = \left(-c_0 + i\frac{z}{|z|}c_1\right)|z|^\alpha$ $(c_0 \geq 0, -\infty < c_1 < \infty, \alpha > 0)$.

PROOF. If $\psi(z) \equiv 0$, then with the choice of $c_0 = c_1 = 0$ the theorem holds, and therefore, we exclude this trivial case from the subsequent discussion. From (2.16.8) it follows that for each positive integer n there exists $a_n > 0$ such that

$$\psi(a_n z) = n\psi(z). \tag{2.16.9}$$

For a positive rational number $r = q/p$ (q, p positive integers), let $a_r = a_q/a_p$. Then, we have

$$\psi(a_r z) = r\psi(z). \tag{2.16.10}$$

If $\psi(\alpha z) = \psi(\beta z)$ holds for some pair $\alpha > \beta > 0$, then letting $\gamma = \beta/\alpha$, we have $\psi(z) = \psi(\gamma z) = \psi(\gamma^2 z) = \cdots = \psi(\gamma^n z) \to \psi(0) = 0$, and hence $\psi(z) \equiv 0$, which is the case we excluded from our consideration. Hence, we see that our choice of a_r above is uniquely determined by the rational number r through (2.16.10). Therefore, we see that, for an arbitrary pair of positive rational numbers r, s the relation $\psi(a_r a_s z) = r\psi(a_s z) = rs\psi(z) = \psi(a_{rs} z)$ holds, and we can conclude that $a_r a_s = a_{rs}$. From (2.16.10) it follows also that

$$\psi(z) = r^n \psi(a_r^{-n} z).$$

We see that if $r \leq 1$, then $a_r \leq 1$ holds, since $r \leq 1$ and $a_r > 1$ imply in view of the equality above that $\psi(z) \to \psi(0) = 0$, which is a contradiction. From this we conclude that if $r \leq s$, then $a_r = a_{r/s} a_s \leq a_s$, which implies that a_r remains bounded when r stays inside of a bounded interval. Furthermore, we see that $a_r \to 1$ when $r \to 1$. This is because if some subsequence of a_r tends to a limit a, then a must be finite by the remark made above, and by letting $r \to 1$ along this subsequence in (2.16.10), we get $\psi(az) = \psi(z)$, which implies that $a = 1$. These facts together tell us that a_r is uniformly continuous with respect to r over bounded intervals, and hence we can define a_t, for any positive real number t, by

$$a_t = \lim_{\substack{r \to t \\ r \in Q}} a_r \quad (Q \text{ is the set of all rational numbers}),$$

and we have

$$\psi(a_t z) = t\psi(z), \quad a_t a_u = a_{tu}, \quad a_t \text{ is continuous in } t.$$

We can show that from the last two conditions stated above we must have $a_t = t^{1/\alpha}$ for some $\alpha > 0$, and for any $b > 0$

$$\psi(bz) = b^\alpha \psi(z)$$

holds. By setting $z = 1$, we get $\psi(b) = b^\alpha \psi(1)$. We also have $\psi(-b) = \overline{\psi(b)} = b^\alpha \overline{\psi(1)}$. Putting these together we obtain that

$$\psi(z) = |z|^\alpha \left(-c_0 + ic_1 \frac{z}{|z|} \right).$$

Since $|\varphi(z)| \leq 1$, we must have $c_0 \geq 0$. □

Since $\varphi(z) = \exp \psi(z)$ is the characteristic function of an infinitely divisible distribution, we can determine a temporally homogeneous additive process x_t, $0 \leq t < \infty$, for which

$$\varphi_{st}(z) = E(e^{iz(x_t - x_s)}) = \exp\{(t-s)\psi(z)\}.$$

If we assume that $\psi(z)$ is written in the form (2.16.7), then the mean of the number of jumps of x_t of magnitude lying in du and occurring in the time interval $[0, t]$ is given by $t \cdot n(du)$. If, for a positive number a, we consider the process $y_t = ax_t$, then this is also an additive process, and

$$\varphi'_{st}(z) \equiv E(e^{iza(x_t - x_s)}) = \exp\{(t-s)\psi(az)\} = \exp\{(t-s)a^\alpha \psi(z)\}.$$

Therefore, the mean of the number of jumps of y_t of magnitude lying in du and occurring in the time interval $[0,t]$ is given by $t \cdot a^\alpha \cdot n(du)$, which equals, since $y_t = a \cdot x_t$, the mean number $t \cdot n(du/a)$ of the jumps of x_t of magnitude lying in du/a occurring in the time interval $[0,t]$. Therefore, we have

$$n(du/a) = a^\alpha \cdot n(du),$$

from which it follows that for $x > 0$

$$n_+(x) = \int_x^\infty n(du) = \int_1^\infty n(x \cdot du) = \int_1^\infty x^{-\alpha} n(du) = x^{-\alpha} n_+(1).$$

We can, therefore, conclude that $n(du) = const \cdot u^{-\alpha-1} du \ (u > 0)$. We get a similar result for $u < 0$ as well. Let us set

$$\begin{aligned} n(du) &= c_+ \cdot u^{-\alpha-1} du \ \ (u > 0) \\ &= c_- |u|^{-\alpha-1} du \ \ (u < 0) \qquad (c_\pm \geq 0). \end{aligned}$$

Then, we have

THEOREM 2.16.3. *Either Φ is a normal distribution or $0 < \alpha < 2$ must be satisfied.*

PROOF. If $c_+ = c_- = 0$ holds, then Φ is a normal distribution. If one of c_+, c_- is positive, then the condition $\int_{-1}^1 u^2 n(du) < \infty$ forces the requirement $0 < \alpha < 2$. □

We separate the argument into the following three cases depending on the value of α: $0 < \alpha < 1$, $1 < \alpha < 2$, $\alpha = 1$.

(a) $0 < \alpha < 1$. In this case, we have

$$\int_{-\infty}^\infty n(du) = \infty, \quad \int_{-1}^1 |u| n(du) < \infty$$

so that

$$\psi(z) = imz - \frac{v}{2} z^2 + c_+ \int_0^\infty (e^{izu} - 1) \frac{du}{u^{\alpha+1}} + c_- \int_{-\infty}^0 (e^{izu} - 1) \frac{du}{|u|^{\alpha+1}}.$$

Terms represented by the integrals in the above are of the order $O(|z|^\alpha)$ (when $z \to 0$ and $z \to \infty$). For instance, if $z > 0$,

$$\int_0^\infty (e^{izu} - 1) \frac{du}{u^{\alpha+1}} = z^\alpha e^{-\pi i \alpha/2} \int_0^\infty (e^{-v} - 1) \frac{dv}{v^{\alpha+1}} \quad (z > 0).$$

Considering that it is of the order $O(|z|^\alpha)$ as $z \to \infty$, we must have $m = v = 0$, and we get

$$\psi(z) = c_+ \int_0^\infty (e^{izu} - 1) \frac{du}{u^{\alpha+1}} + c_- \int_{-\infty}^0 (e^{izu} - 1) \frac{du}{|u|^{\alpha+1}}.$$

Therefore, in this case a sample process of the corresponding additive process x_t is with probability 1 a purely discontinuous function varying only with jumps. Furthermore, except for the special case of $c_+ = c_- = 0$, the number of jumps is infinite with probability 1. In Theorem 2.16.2 we showed that $c_0 \geq 0$, but if $c_0 = 0$, then we have $|\varphi(z)| = 1$, so that Φ is a δ-distribution, and hence $\psi(z) = imz$ (m is a constant), but since $0 < \alpha < 1$, it must be the case that $c_1 = m = 0$, which implies that Φ must be the unit distribution. As $|\varphi(z)| = \exp(-c_0|z|^\alpha)$ $(0 < \alpha < 1)$, $\varphi(z)$ belongs to both $L_1(R^1)$ and $L_2(R^1)$, which implies that its Fourier transform is a continuous function. This in turn shows that the distribution Φ corresponding to φ has a continuous density function. It is a very interesting fact that while the sample process is with probability 1 a purely discontinuous function, the distribution of the process has a continuous density. This illustrates that discontinuous quantities may get averaged out to become continuous.

(b) $1 < \alpha < 2$. In this case we have

$$\int_{-\infty}^{\infty} n(du) = \infty, \quad \int_{-\infty}^{\infty} un(du) = \infty, \quad \int_{|u|>1} |u| n(du) < \infty.$$

Consequently,

$$\psi(z) = imz - \frac{v}{2}z^2 + c_+ \int_0^{\infty} (e^{izu} - 1 - izu)\frac{du}{u^{\alpha+1}} + c_- \int_{-\infty}^{0} (e^{izu} - 1 - izu)\frac{du}{|u|^{\alpha+1}}.$$

Terms represented by integrals in the above are of the order $O(|z|^\alpha)$, and hence by considering the order when $z \to 0$ and $z \to \infty$, we see that $m = v = 0$ must hold, and we get

$$\psi(z) = c_+ \int_0^{\infty} (e^{izu} - 1 - izu)\frac{du}{u^{\alpha+1}} + c_- \int_{-\infty}^{0} (e^{izu} - 1 - izu)\frac{du}{|u|^{\alpha+1}}.$$

The number of jumps and the sum of the absolute values of the magnitude of jumps for the corresponding additive process are both infinite with probability 1. However, for an arbitrary β ($> \alpha$), the sum of the β-th powers of the absolute value of the magnitude of jumps

$$\sum_{0<\tau<t} |x_{\tau+0} - x_{\tau-0}|^\beta$$

is finite with probability 1. This can be shown in the following way. Since the number of jumps of which the absolute value of the magnitude exceeds 1 is finite with probability 1, it is enough to consider the number of jumps of which the absolute value of the magnitude is ≤ 1, and the mean number of such is

$$\int_0^t \int_{-1}^1 |u|^\beta \frac{du}{u^{\alpha+1}} d\tau < \infty,$$

and hence the number of such jumps must also be finite with probability 1.

(c) $\alpha = 1$. In this case

$$\psi(z) = ic_1 z - c_0|z|$$

and since

$$\int_{-\infty}^{\infty} \left(e^{izu} - 1 - \frac{izu}{1+u^2}\right)\frac{du}{u^2} = 2\int_0^{\infty} (\cos zu - 1)\frac{du}{u^2} = -\pi|z|,$$

we get

$$\psi(z) = ic_1 z + \frac{c_0}{\pi}\int_{-\infty}^{\infty} \left(e^{izu} - 1 - \frac{izu}{1+u^2}\right)\frac{du}{u^2}.$$

Therefore, the corresponding distribution is a Cauchy distribution. For this reason, the corresponding additive process is called a **Cauchy process**.

Additive processes corresponding with stable distributions such as those we have seen above are called **stable processes**. Stable processes consist of the cases (a), (b) discussed above, Cauchy processes of the case (c), Wiener processes (cf. Theorem 2.16.3), and their degenerate case $x_t = m \cdot t$ (m is a constant).

CHAPTER 3

Stationary Processes

3.1. Definition of Stationary Process

A stationary process is a stochastic process which describes phenomena which stay stationary under the transition of time. There are two ways of defining the notion of stationarity, strong and weak. Let x_t, $t \in T = (-\infty, \infty)$, be a stochastic process, and let

$$m(t) = E(x_t), \quad v(t,s) = E((x_t - E(x_t))(x_s - E(x_s))),$$
$$\Phi_{t_1,t_2,\cdots,t_n}(E) = P\{\omega/(x_{t_1}, x_{t_2}, \cdots, x_{t_n}) \in E\}.$$

x_t is called a **weakly stationary stochastic process** if for every choice of t, s, and h,

$$m(t+h) = m(t), \quad v(t+h, s+h) = v(t,s)$$

are satisfied. In this case, $m(t)$ becomes a constant (m), and $v(t,s)$ becomes a function $v(t-s)$ of $t-s$.

If for every choice of n, $\{t_i\}$,

$$\Phi_{t_1+h,t_2+h,\cdots,t_n+h} = \Phi_{t_1,t_2,\cdots,t_n}$$

holds, then x_t is called a **strongly stationary stochastic process**.

THEOREM 3.1.1. *If x_t, $t \in T$, is strongly stationary and $E(|x_0|^2) < \infty$, then x_t, $t \in T$, is weakly stationary.*

PROOF. From the strong stationarity it follows that $E(|x_t|^2) = E(|x_0|^2) < \infty$. Therefore, both $m(t), v(t,s)$ are well defined and finite. Furthermore, since

$$m(t) = \int \xi \Phi_t(d\xi), \quad v(t,s) = \int\int (\xi - m(t))(\eta - m(s))\Phi_{ts}(d(\xi,\eta)),$$

the weak stationarity follows from the strong stationarity. □

The converse of this fact does not necessarily hold, but for a normal stochastic process the following theorem holds.

THEOREM 3.1.2. *If a normal stochastic process x_t is weakly stationary, then it is strongly stationary.*

PROOF. For an arbitrary choice of $t_1 < t_2 < \cdots < t_n$, if we let

$$M = (m(t_i)), \quad V = (v(t_i, t_j)),$$

then from the assumption of normality it follows that $\Phi_{t_1,\cdots,t_n}$ is given by $N(\bullet; M, V)$. From the assumption of weak stationarity, we see that M and V stay the same when t_i is replaced by $t_i + h$ for each i. Therefore, $\Phi_{t_1+h,t_2+h,\cdots,t_n+h} = \Phi_{t_1,t_2,\cdots,t_n}$, that is, x_t is strongly stationary. □

If t varies over the set of integers $\{\cdots, -3, -2, -1, 0, 1, 2, 3, \cdots\}$ instead of $(-\infty, \infty)$, the same results as above hold. We say in this case x_t is a **stationary random sequence**.

When $x_t(\omega)$ is a complex-valued stochastic process, namely when $x_t(\omega)$ takes complex values, we can also define stationarity. The only difference from the real-valued case is that the definition of $v(t, s)$ is given by

$$v(t, s) = E((x_t - m(t))\overline{(x_s - m(s))}) \quad (\bar{\xi} = \text{complex conjugate of } \xi).$$

Theorem 3.1.1 remains valid as it is. What corresponds to Theorem 3.1.2 is also valid, but in order to state the result precisely we need to introduce the notion of a complex normal distribution (cf. §3.6, §3.7). In the sequel, we consider complex-valued stationary processes unless stated otherwise.

3.2. Preliminary Material Related to Investigations of Stationary Processes

In this section we put together the preliminary material necessary for the study of stationary processes.

We already explained the following theorem in §1.4, where we dealt with characteristic functions, but we state it here again in a slightly different form.

THEOREM 3.2.1. **Bochner's Theorem.** *If a complex-valued function $\varphi(t)$ defined for $-\infty < t < \infty$ satisfies the following two properties:*

(1) *positive definiteness: $\sum_{ij} \varphi(t_i - t_j)\xi_i\bar{\xi}_j \geq 0$,*
(2) *continuity: when $t \to 0$, then $\varphi(t) \to \varphi(0)$,*

then there exists a bounded right continuous increasing function $F(\lambda)$ such that

$$\varphi(t) = \int_{R^1} e^{-i2\pi t\lambda} dF(\lambda), \quad F(-\infty) = 0.$$

We inserted 2π in the exponent above for the sake of convenience in subsequent discussions.

THEOREM 3.2.2. **Stone's Theorem.** *Suppose a 1-parameter family U_t, $-\infty < t < \infty$, of unitary operators on a Hilbert space $\boldsymbol{H}$ satisfying the following two conditions is given:*

(1) *continuity: $(U_t f, g)$ is continuous as a function of t (or it is sufficient to assume the measurability in t),*
(2) *group property: $U_t U_s = U_{t+s}$.*

Then U_t has the **spectral decomposition***:*

$$U_t = \int e^{-i2\pi\lambda t} dE(\lambda).$$

Also, if U is a unitary operator, then it has the spectral decomposition given by

$$U = \int e^{-i2\pi\lambda} dE(\lambda).$$

THEOREM 3.2.3. **Ergodic Theorem.** *Let $\Omega(\boldsymbol{B}, P)$ be a probability space, and let S be a 1 to 1 measure preserving transformation of Ω onto itself. (Here, by S*

being measure preserving we mean that if E is a measurable set, so are both SE and $S^{-1}E$ and $P(SE) = P(S^{-1}E) = P(E)$ holds.) Then, for $f \in L^1(\Omega)$,

$$f^*(\omega) = \lim_{\substack{n\to\infty \\ m\to-\infty}} \frac{1}{n-m}[f(S^{m+1}\omega) + f(S^{m+2}\omega) + \cdots + f(S^n\omega)]$$

exists for almost all ω, and furthermore, $f^*(S\omega) = f^*(\omega)$ holds a.e.

This theorem is called the **Individual Ergodic Theorem** of G. D. Birkhoff.

We assume that the readers are familiar with the theory of **L. Schwartz' distribution**. We will consider in addition to the usual complex-valued Schwartz' distributions those taking values in a Hilbert space as well. The definition of the latter is given in exactly the same way as in the usual case.

If $\Omega(\boldsymbol{B}, P)$ is a probability space, $\boldsymbol{H} = L^2(\Omega)$ can be regarded as a Hilbert space in the usual way. Some of the probabilistic concepts defined on $\Omega(\boldsymbol{B}, P)$ can be restated using Hilbert space terminologies as follows:

$$E(x) = (x, 1),$$
$$E(x \cdot \overline{y}) = (x, y), \text{ in particular, } E(|x|^2) = \|x\|^2,$$
$$x_n \to x \text{ (mean square convergence)} \iff \|x_n - x\| \to 0.$$

If x_t ($\in \boldsymbol{H}$) is continuous (in the sense of norm topology) and bounded in norm, and if $f(t) \in L^1(R^1)$, then $\int f(t)x_t dt$ can be defined in the sense of norm convergence. The integral over a t-interval can also be defined in the same way. Next, let us define an integral of the form

$$I(f) = \int f(t) dy_t.$$

A particularly important case arises when y_t has **orthogonal increments**, namely when $y_{t_2} - y_{t_1}$ and $y_{s_2} - y_{s_1}$ become orthogonal whenever intervals $(t_1, t_2]$ and $(s_1, s_2]$ are disjoint. In this case, a monotone increasing function $F(t)$ (unique up to an additive constant) can be determined for which

$$F(t) - F(s) = \|y_t - y_s\|^2 \quad (t > s)$$

holds. If $F(t)$ is assumed to be right continuous, then y_t is also right continuous. Let us denote by the same F the Lebesgue-Stieltjes measure determined by the function F. That is,

$$F(E) = \int_E dF(t).$$

Let us show that for $f \in L^2(R^1, F)$, the integral $I(f)$ mentioned above can be defined. Denote by $\boldsymbol{J}$ the set of all step functions with bounded support, and let us define $I(f)$ for $f \in \boldsymbol{J}$ in the usual manner. Then we see that $I(f) \in \boldsymbol{H}$ and

$$\|I(f)\| = \|f\| \quad (\|f\| \text{ here denotes the norm in } L^2(R^1, F))$$

holds. Therefore, we can extend by the usual method the definition of $I(f)$ from $\boldsymbol{J}$ to its closure $\overline{\boldsymbol{J}} = L^2(R^1, F)$ to get a linear operator from $L^2(R^1, F)$ to $\boldsymbol{H}$. This extension is the desired integral $I(f)$, and it satisfies

$$I(\alpha f + \beta g) = \alpha I(f) + \beta I(g),$$
$$(I(f), I(g)) = (f, g).$$

3.3. Spectral Decomposition of Weakly Stationary Processes

Let x_t, $-\infty < t < \infty$, be a weakly stationary process, and let

$$E(x_t) = m, \quad E((x_t - m)\overline{(x_s - m)}) = v(t - s). \tag{3.3.1}$$

Since $v(0) > 0$ holds except for the trivial special case $x_t(\omega) \equiv m$, which we exclude, we may assume, without loss of generality, by considering $(x_t - m)/\sqrt{v(0)}$ that

$$E(x_t) = 0, \quad E(x_t\overline{x}_s) = v(t - s), \quad E(|x_t|^2) = v(0) = 1 \tag{3.3.2}$$

hold.

x_t, $-\infty < t < \infty$, can be regarded as a curve lying in the Hilbert space $\boldsymbol{H} = L^2(\Omega)$. Furthermore, since we have from (3.3.2)

$$(x_t, 1) = 0, \quad (x_t, x_s) = v(t - s), \quad \|x_t\| = 1,$$

we see that this curve lies in the orthocomplement $\boldsymbol{H}' = \{y/y \perp 1\}$ of $\{1\}$ in $\boldsymbol{H}$, and also on the unit sphere of $\boldsymbol{H}$. $v(t-s)$ represents the cosine of the angle between the two vectors x_t and x_s of $\boldsymbol{H}$. This quantity is sometimes called the **correlation coefficient** of x_t, x_s.

In the sequel, we assume further: the **norm continuity** of x_t. Namely,

$$\lim_{t \to s} \|x_t - x_s\| = 0. \tag{3.3.3}$$

As we have seen for Poisson processes in the preceding chapter, this norm continuity does not necessarily imply the continuity of sample processes. However, since

$$P\{\omega/|x_t - x_s| > \epsilon\} \leq \|x_t - x_s\|^2/\epsilon^2$$

holds, the continuity in probability does follow from the norm continuity.

THEOREM 3.3.1 (A. Khinchin). *The spectral decomposition of $v(t)$ is valid:*

$$v(t) = \int e^{-i2\pi\lambda t} dF(\lambda), \quad F(-\infty) = 0. \tag{3.3.4}$$

Here, $F(\lambda)$ is a bounded right continuous increasing function determined by $v(t)$, and is called the spectral function of $v(t)$.

PROOF. From the definition of $v(t)$ it follows that

$$\sum_{ij} v(t_i - t_j)\xi_i\overline{\xi}_j = \|\sum \xi_i x(t_i)\|^2 \geq 0.$$

Also by the norm continuity we have

$$v(t) = (x_t, x_0) \to (x_0, x_0) = v(0) \quad (t \to 0).$$

Therefore, by Bochner's Theorem, we see that the desired function F is determined. □

THEOREM 3.3.2 (A. Kolmogorov). *The spectral decomposition of x_t can be obtained as follows:*

$$x_t = \int e^{-i2\pi\lambda t} dy_\lambda, \quad y_{-\infty} = 0, \tag{3.3.5}$$

where y_λ has orthogonal increments and

$$\|y_\lambda\|^2 = F(\lambda). \tag{3.3.6}$$

y_λ is determined completely by x_t.

PROOF. Let us denote by $\boldsymbol{A}$ the subspace of $\boldsymbol{H}$ formed by linear combinations of $\{x_t\}$ and let $\boldsymbol{H}_0$ be the closure $\overline{\boldsymbol{A}}$. Define a transformation group U_t, $-\infty < t < \infty$, on $\boldsymbol{A}$ by

$$U_t(\sum a_i x_{t_i}) = \sum a_i x_{t_i+t}.$$

To see that this definition is well defined it is enough to note that

$$\sum a_i x_{t_i} = 0 \Rightarrow \sum a_i x_{t_i+t} = 0.$$

But this follows obviously from

$$\| \sum a_i x_{t_i} \|^2 = \sum a_i \overline{a_j} v(t_i - t_j) = \| \sum a_i x_{t_i+t} \|^2.$$

Furthermore, the identity above shows also that U_t is an **isometric transformation** from $\boldsymbol{A}$ to $\boldsymbol{A}$. Consequently, we can extend U_t to be an isometric transformation of $\boldsymbol{H}_0$ to $\boldsymbol{H}_0$. In addition, since $U_t U_s = U_{t+s}$ is satisfied in $\boldsymbol{A}$, the same property holds in $\boldsymbol{H}_0$ as well. Since $U_t f$ is norm continuous in t when $f \in \boldsymbol{A}$, we can show by using the isometric property of U_t that it is also norm continuous in t when $f \in \boldsymbol{H}_0$. These facts allow us to use Stone's Theorem to obtain the spectral decomposition of U_t, so that we can write

$$x_t = U_t x_0 = \int e^{-i2\pi\lambda t} d(E(\lambda)x_0) = \int e^{-i2\pi\lambda t} dy_\lambda.$$

From the properties of the spectral decomposition it follows that y_λ has orthogonal increments, and the function $G(\lambda)$ given by

$$G(\lambda) = \|y_\lambda\|^2$$

becomes a bounded right continuous increasing function. Furthermore, again from the properties of the spectral decomposition, we get

$$v(t) = (x_t, x_0) = \int e^{i2\pi\lambda t} dG(\lambda), \quad G(-\infty) = 0.$$

By comparing this with (3.3.4), we get $F = G$, and therefore, (3.3.6) is satisfied.

Next, we will show that y_λ is determined completely by x_t through (3.3.5). So, let us suppose that both y_λ and y'_λ satisfy (3.3.5). If we denote by $\hat{f}$ the Fourier transform of $f \in L^1(R^1)$, then since both y_λ and y'_λ satisfy (3.3.5), we have

$$\int \hat{f}(\lambda) dy_\lambda = \int \hat{f}(\lambda) dy'_\lambda = \int f(t) x_t dt.$$

If $g(\lambda)$ is a continuous function with compact support, then $g(\lambda)$ can be approximated uniformly on R^1 by such functions $\hat{f}$, and therefore, we have

$$\int g(\lambda) dy_\lambda = \int g(\lambda) dy'_\lambda.$$

Suppose for a fixed $\mu \in R^1$, we denote by $c(\lambda)$ the indicator function of the set $(-\infty, \mu]$. Then for any given ϵ, we can choose $g(\lambda)$, a continuous function with compact support, for which

$$\int |c(\lambda) - g(\lambda)|^2 dF(\lambda) < \epsilon^2,$$

so that

$$\| \int c(\lambda) dy_\lambda - \int g(\lambda) dy_\lambda \|^2 = \int |c(\lambda) - g(\lambda)|^2 dF(\lambda) < \epsilon^2.$$

Similarly, we get

$$\| \int c(\lambda) dy'_\lambda - \int g(\lambda) dy'_\lambda \|^2 < \epsilon^2.$$

Consequently, we get

$$\| \int c(\lambda) dy_\lambda - \int c(\lambda) dy'_\lambda \| < 2\epsilon.$$

Letting $\epsilon \downarrow 0$, we obtain

$$\int c(\lambda) dy_\lambda = \int c(\lambda) dy'_\lambda, \text{ which means that } y_\mu = y'_\mu.$$

This completes the proof of the theorem. □

EXAMPLE 3.3.1. When $F(\lambda)$ is, in particular, a purely discontinuous function, then a sequence $\{\lambda_n\}$ of real numbers and a sequence $\{a_n\}$ of positive numbers with $\sum a_n < \infty$ exist so that $F(\lambda)$ can be written as

$$F(\lambda) = \sum_{\lambda_n \leq \lambda} a_n.$$

In this case,

$$v(t) = \int e^{-i2\pi\lambda t} dF(\lambda) = \sum a_n e^{-i2\pi\lambda_n t}$$

becomes an almost periodic function. The spectral decomposition of the corresponding x_t is given by

$$x_t = \sum y_n e^{-i2\pi\lambda_n t},$$

where $\{y_n\}$ is an orthogonal sequence and $\|y_n\|^2 = a_n$. In fact, we can define y_n to be equal to $y_{\lambda_n+0} - y_{\lambda_n-0}$.

3.4. Spectral Decomposition of Sample Processes of Weakly Stationary Processes

Let x_t, $-\infty < t < \infty$, be a weakly stationary process, and suppose the conditions (3.3.2), (3.3.3) in the preceding section are satisfied. Then by Kolmogorov's Spectral Decomposition Theorem, we have the spectral decomposition

$$x_t = \int e^{-i2\pi\lambda t} dy_\lambda, \quad y_{-\infty} = 0. \tag{3.4.1}$$

Let us consider this relationship in terms of sample processes. The integral on the right-hand side of the equation above is defined using the norm convergence in $L^2(\Omega)$, and therefore, this identity cannot be regarded as giving a relationship concerning the values of sample processes evaluated at each ω. Let us first observe that by using the facts

$$\|x_t - x_s\| \to 0 \ (t \to s), \quad \|y_\lambda - y_\mu\| \to 0 \ (\lambda \downarrow \mu), \tag{3.4.2}$$

the following theorem can be proved.

THEOREM 3.4.1. *There exist processes x_t^* and y_λ^* which are measurable in two variables (t, ω) and (λ, ω), respectively, and which are equivalent with x_t and y_λ, respectively, in the weak sense (cf. §2.8).*

We omit the proof of this theorem. Since it is clear that the relation (3.4.1) is satisfied between x_t^* and y_λ^*, we will assume in the sequel that x_t, y_λ are measurable in two variables. From the fact that $x_t(\omega)$ (or $y_\lambda(\omega)$) is measurable in two variables (t,ω) (or (λ,ω)) it follows that for almost all ω, x_t (or y_λ) is measurable in t (or in λ).

$x_t(\omega)$ is a **slowly increasing function** of t for almost all ω (cf. Schwartz' distribution theory). This is true since

$$E\left\{\left(\int \frac{|x_t|}{1+t^2}dt\right)^2\right\} \le E\left\{\int \frac{|x_t|^2}{1+t^2}dt \int \frac{dt}{1+t^2}\right\} = \left(\int \frac{dt}{1+t^2}\right)^2 < \infty,$$

and therefore,

$$\int \frac{|x_t(\omega)|}{1+t^2}dt < \infty$$

for almost all ω and this implies that for almost all ω, $x_t(\omega)$ is a slowly increasing (Schwartz') distribution (in fact, slowly increasing function). Similarly, $y_\lambda(\omega)$ is a slowly increasing function of λ for almost all ω, since

$$E\left\{\left(\int \frac{|y_\lambda|}{1+\lambda^2}d\lambda\right)^2\right\} \le E\left\{\int \frac{|y_\lambda|^2}{1+\lambda^2}d\lambda \int \frac{d\lambda}{1+\lambda^2}\right\} \le F(\infty)\left(\int \frac{d\lambda}{1+\lambda^2}\right)^2 < \infty.$$

From this it follows also that the derivative Dy_λ of y_λ in the sense of Schwartz' distribution is also a slowly increasing Schwartz' distribution.

THEOREM 3.4.2. *For almost all ω, $x_t(\omega)$, $-\infty < t < \infty$, is the Fourier transform of the Schwartz' distribution $Dy_\lambda(\omega)$. Namely,*

$$x(\phi) = Dy_\lambda(\mathfrak{F}\phi) \tag{3.4.3}$$

holds for every test function ϕ. Here

$$\mathfrak{F}\phi(\lambda) = \int e^{-i2\pi\lambda t}\phi(t)\,dt.$$

PROOF. Since

$$E\left[\int \frac{|x_t(\omega)|^2}{1+t^2}dt\right] = \int \frac{1}{1+t^2}dt < \infty$$

holds, we have $P(\Omega_1) = 1$ if we denote

$$\Omega_1 = \left\{\omega \Big/ \int \frac{|x_t(\omega)|^2}{1+t^2}dt < \infty\right\}.$$

If for $\omega \in \Omega_1$, we define

$$z_\lambda(\omega) = \int_{-1}^{1} x_t \frac{e^{i2\pi\lambda t}-1}{i2\pi t}dt + \operatorname*{l.i.m.}_{a\to\infty}{}^{\dagger}\left(\int_1^a + \int_{-a}^{-1}\right) x_t \frac{e^{i2\pi\lambda t}}{i2\pi t}dt \tag{3.4.4}$$

$$\left(\operatorname*{l.i.m.}_{a\to\infty} \eta_a(\omega) = \eta(\omega) \Leftrightarrow \lim_{a\to\infty} \int |\eta_a(\omega) - \eta(\omega)|^2 P(d\omega) = 0\right),$$

then we get

$$x(\phi) = \mathfrak{F}Dz_\lambda(\phi), \quad \phi \in \mathfrak{D}. \tag{3.4.5}$$

† l.i.m. = limit in the mean.

Namely, if $\omega \in \Omega_1$, then x and $\mathfrak{F}\ Dz_\lambda$ become the same element of $\mathfrak{D}'$, and since $P(\Omega_1) = 1$, this implies further that x and $\mathfrak{F}\ Dz_\lambda$ coincide as elements of $\mathfrak{D}'_{\boldsymbol{H}}$, $\boldsymbol{H} = L^2(\Omega)$. Therefore, in $\boldsymbol{H}$, we have

$$x(\phi) = \int \mathfrak{F}\phi(\lambda)\,dy_\lambda = -\int (\mathfrak{F}\phi(\lambda))' y_\lambda\, d\lambda,$$

$$x(\phi) = \mathfrak{F}Dz_\lambda(\phi) = -\int (\mathfrak{F}\phi(\lambda))' z_\lambda\, d\lambda.$$

If we set, for an arbitrary rapidly decreasing function ψ, $\phi = \mathfrak{F}^{-1}\psi$, then

$$\int \psi' y_\lambda d\lambda = \int \psi' z_\lambda d\lambda, \text{ and hence } Dy_\lambda = Dz_\lambda$$

from which we deduce that for almost every λ, $y_\lambda = z_\lambda + c$, $c \in \boldsymbol{H}$, holds. Since we can take $z_\lambda - c$ in place of z_λ in (3.4.5), we can conclude that $y_\lambda = z_\lambda$ holds in $\boldsymbol{H}$ for almost all λ. From (3.4.4) we see that for almost every λ, we can take a sequence $a_n \to \infty$ such that

$$z_\lambda(\omega) = \int_{-1}^{1} x_t \frac{e^{i2\pi\lambda t} - 1}{i2\pi t} dt + \lim_{a_n\to\infty} \left(\int_1^{a_n} + \int_{-a_n}^{-1} \right) x_t \frac{e^{i2\pi\lambda t}}{i2\pi t} dt.$$

Therefore, by using the fact that $x_t(\omega)$ is measurable in the two variables (t, ω), we can take $z_\lambda(\omega)$ to be measurable in the two variables (λ, ω). As we remarked earlier, $y_\lambda(\omega)$ can be assumed to be measurable in the two variables (λ, ω). We saw also that for almost all λ, $y_\lambda = z_\lambda$ holds for almost all ω. Consequently, by Fubini's Theorem, we have for almost all ω, $y_\lambda(\omega) = z_\lambda(\omega)$ holds for almost all λ. Thus we see that (3.4.5) can be restated as $x(\phi) = \mathfrak{F}Dy_\lambda(\phi)$. $\square$

(3.4.3) can be rewritten in the following form à là N. Wiener:

(3.4.6) $$x_t = \lim_{\epsilon\to 0} \underset{a\to\infty}{\text{l.i.m.}} \int_{-a}^{a} e^{-i2\pi\lambda t} \frac{y_{\lambda+\epsilon} - y_{\lambda-\epsilon}}{2\epsilon} d\lambda.$$

3.5. Ergodic Theorem Concerning Strongly Stationary Processes

Let x_t, $-\infty < t < \infty$, be a measurable strongly stationary process. It is clear that

$$E(|x_t|) = E(|x_0|)$$

holds, and we assume that this value is finite. The assumption of measurability above means that $x_t(\omega)$ is measurable in the two variables (t, ω), from which it follows that for almost all ω, $x_t(\omega)$ is a measurable function of t. Furthermore, since for $-\infty < a < b < \infty$

$$E\left\{\int_a^b |x_t| dt\right\} = \int_a^b E(|x_t|)dt = E(|x_0|)(b-a) < \infty$$

is valid, for an arbitrary finite interval (a, b),

$$\int_a^b |x_t(\omega)| dt < \infty$$

holds for almost all ω. If this integral is finite whenever a, b are integers, then it is finite for an arbitrary finite interval. Since the set of all integral pairs $a < b$ is a countable set, we can choose a single exceptional set of ω's of probability 0, outside of which the integral above over (a, b) with a, b being integers is finite, and therefore, the integral above over an arbitrary finite interval (a, b) is finite for all ω outside a single exceptional set of probability 0.

THEOREM 3.5.1. *If $E(|x_0|) < \infty$, then for almost all ω,*

$$x^*(\omega) = \lim_{A\to\infty} \frac{1}{2A} \int_{-A}^{A} x_t(\omega)dt$$

exists. This limit is called the **sample mean**.

PROOF. We start with the following lemma.

LEMMA 3.5.1. *If $\{x_n\}$ is a stationary sequence for which $E(|x_0|) < \infty$, then*

$$x^*(\omega) = \lim_{\substack{n\to\infty \\ m\to-\infty}} \frac{1}{n-m} \sum_{k=m+1}^{n} x_k(\omega)$$

exists for almost all ω

PROOF OF THE LEMMA. Let $Z = \{\cdots, -3, -2, -1, 0, 1, 2, \cdots\}$, and equip the measurable space $R^Z(\boldsymbol{B}^Z)$ with the distribution Φ of the random vector $\boldsymbol{x} = \prod_k x_k$ to form a probability space $R^Z(\boldsymbol{B}^Z, \Phi)$. Consider the 1 to 1 mapping T of R^Z onto itself defined by

$$T : \prod_k \xi_k \to \prod_k \xi_{k+1}.$$

Then T is a measure preserving transformation of the space $R^Z(\boldsymbol{B}^Z, \Phi)$. If we consider a real-valued function f on R^Z defined by $f : \Xi = \prod_k \xi_k \to \xi_0$, then $f \in L^1(R^Z)$. Therefore, for almost all (with respect to Φ) Ξ,

$$f^*(\Xi) = \lim_{\substack{n\to\infty \\ m\to-\infty}} \sum_{k=m+1}^{n} f(T^k\Xi)$$

exists. Consequently, we can conclude that

$$f^*(\boldsymbol{x}(\omega)) = \lim_{\substack{n\to\infty \\ m\to-\infty}} \frac{1}{n-m} \sum_{k=m+1}^{n} x_k(\omega)$$

exists for almost all ω, proving the lemma. □

Let us now prove the theorem. If we define

$$y_n = \int_n^{n+1} x_t dt, \quad n = \cdots, -3, -2, -1, 0, 1, 2, \cdots,$$

then y_n becomes a stationary sequence. (The proof of this fact is rather cumbersome, so we omit it.) From the lemma above it follows that

$$x^* = \lim_{n\to\infty} \frac{1}{2n} \int_{-n}^{n} x_t dt = \lim_{n\to\infty} \frac{1}{2n} \sum_{k=-n}^{n-1} y_k$$

exists for almost all ω. Next, we observe that if $n < A < n+1$, then

$$\frac{1}{2A} \int_{-A}^{A} x_t dt = \frac{n}{A} \cdot \frac{1}{2n} \int_{-n}^{n} x_t dt + \frac{1}{2A} \int_{-A}^{-n} x_t dt + \frac{1}{2A} \int_{n}^{A} x_t dt.$$

Therefore, in order to prove the theorem, it suffices to show that the last two terms on the right-hand side of the above identity tend to 0 as $A \to \infty$. If we apply the arguments above to $|x_t|$, we get

$$\frac{1}{2n}\int_{-(n+1)}^{n+1}|x_t|dt = \frac{n+1}{n}\cdot\frac{1}{2(n+1)}\int_{-(n+1)}^{n+1}|x_t|dt \to |x|^*,$$

$$\frac{1}{2n}\int_{-n}^{n}|x_t|dt \to |x|^*.$$

Hence by considering the difference, we obtain

$$\frac{1}{2n}\int_{-(n+1)}^{-n}|x_t|dt + \frac{1}{2n}\int_{n}^{n+1}|x_t|dt \to 0,$$

from which it follows that

$$\left|\frac{1}{2A}\int_{-A}^{-n}x_t dt + \frac{1}{2A}\int_{n}^{A}x_t dt\right| \le \frac{1}{2n}\int_{-(n+1)}^{-n}|x_t|dt + \frac{1}{2n}\int_{n}^{n+1}|x_t|dt \to 0. \qquad \square$$

If $f(\xi_1, \xi_2, \cdots, \xi_n)$ is a Baire function of n variables, then

$$y_t = f(x_{t_1+t}, x_{t_2+t}, \cdots, x_{t_n+t})$$

becomes also a separable strongly stationary process, and therefore, if

$$E(|y_0|) = E(|f(x_{t_1}, x_{t_2}, \cdots, x_{t_n})|) < \infty,$$

then we can apply the preceding theorem to y_t and show the existence of the limit

$$\lim_{A\to\infty}\frac{1}{2A}\int_{-A}^{A}f(x_{t_1+t}, x_{t_2+t}, \cdots, x_{t_n+t})dt.$$

If, in particular, $E(|x_0|^2) < \infty$, and hence $E(|x_t\overline{x}_s|) < \infty$, then

$$\lim_{A\to\infty}\frac{1}{2A}\int_{-A}^{A}x_{t+\sigma}\overline{x}_{s+\sigma}d\sigma$$

exists, and hence we know that

$$\begin{aligned} v^*(t,s) &= \lim_{A\to\infty}\frac{1}{2A}\int_{-A}^{A}(x_{t+\sigma}-x^*)(\overline{x}_{s+\sigma}-x^*)d\sigma \\ &= \lim_{A\to\infty}\frac{1}{2A}\int_{-A}^{A}x_{t+\sigma}\overline{x}_{s+\sigma}d\sigma - |x^*|^2 \end{aligned}$$

exists also. $V^* = (v^*(t,s))$ is called the sample covariance matrix. It is clear that

$$v^*(t,s) = v^*(t-s) \ (\equiv v^*(t-s,0))$$

holds. Since

$$\sum_{i,j}v^*(t_i-t_j)\xi_i\overline{\xi}_j = \lim_{A\to\infty}\frac{1}{2A}\int_{-A}^{A}\left|\sum\xi_i(x_{t_i+\sigma}-x^*)\right|^2 d\sigma \ge 0$$

holds, we can write

$$v^*(t) = \int e^{-i2\pi\lambda t}dS^*(\lambda) \ \ (\text{with } S^*(-\infty) = 0).$$

$S^*(\lambda)$ is called the sample spectral function. It is clear that

$$E(x^*) = 0 = E(x_0)$$

holds. Furthermore, we have

$$E(v^*(t)) = v(t) - E(|x^*|^2) \le v(t),$$
$$E(S^*(\lambda)) = F(\lambda) - E(|x^*|^2)H(\lambda) \le F(\lambda)$$
$$(H(\lambda) = 0\ (\lambda < 0),\ = 1\ (\lambda \ge 0)).$$

If, in particular, $E(|x^*|^2) = 0$, that is, if $x^* = 0$, then the equalities $E(v^*(t)) = v(t)$, $E(S^*(\lambda)) = F(\lambda)$ hold. The fact that $x^* = 0$ is equivalent to the condition $F(+0) = F(-0)$ can be seen from

$$E(|x^*|^2) = \lim_{A\to\infty} \left(\frac{1}{2A}\right)^2 \int_{-A}^{A}\int_{-A}^{A} v(t-s)dtds = F(+0) - F(-0).$$

As we saw in the preceding section, we have the following identity in the sense of Schwartz' distribution theory:

$$x_t = \int e^{-i2\pi\lambda t} dy_\lambda = \mathfrak{F}(Dy_\lambda).$$

If we use this identity and compute $v^*(t)$ formally, then we can deduce

$$\begin{aligned} v^*(t) &= \lim_{A\to\infty} \frac{1}{2A}\int_{-A}^{A} \iint e^{-i2\pi[\lambda(t+s)-\mu s]} ds dy_\lambda \overline{dy_\mu} \\ &= \iint e^{-i2\pi\lambda t} \lim_{A\to\infty}\frac{1}{2A}\int_{-A}^{A} e^{-i2\pi s(\lambda-\mu)} ds dy_\lambda \overline{dy_\mu} \\ &= \iint e^{-i2\pi\lambda t}\delta_{\lambda\mu} dy_\lambda \overline{dy_\mu}, \quad \text{where } \delta_{\lambda\mu} = 1\ (\lambda=\mu),\ = 0\ (\lambda\ne\mu), \\ &= \int e^{-i2\pi\lambda t}|dy_\lambda|^2. \end{aligned}$$

This formal result leads us to a very interesting symbolic relationship:

$$dS^*(\lambda) = |dy_\lambda|^2,$$

which can be made rigorous by using the generalized harmonic analysis of N. Wiener. A more precise meaning of the last formal identity written above is that for an arbitrary bounded continuous function $f(\lambda)$ the identity

$$\int f(\lambda) dS^*(\lambda) = \lim_{\epsilon\downarrow 0}\int f(\lambda)\frac{|y_{\lambda+\epsilon} - y_{\lambda-\epsilon}|^2}{2\epsilon}d\lambda$$

holds.

3.6. Complex Normal System

We define x_α, $\alpha \in A$, to be a real normal system if the distribution of the random vector $\boldsymbol{x} = \prod_\alpha x_\alpha$ is given by a normal distribution on $R^A(\boldsymbol{B}^A)$. Wiener processes and (real) normal processes are examples of real normal systems. The definition given above is equivalent to the following: for arbitrary $\alpha_\nu \in A$ and $z_\nu \in R^1$

$$E\{\exp(i\sum z_\nu x_{\alpha_\nu})\} = \exp\left\{i\sum z_\nu m(\alpha_\nu) - \frac{1}{2}\sum z_\mu z_\nu v(\alpha_\mu, \alpha_\nu)\right\}$$
$$(m(\alpha) = E(x_\alpha),\ v(\alpha,\beta) = E\{(x_\alpha - m(\alpha))(x_\beta - m(\beta))\})$$

hold. If the expectation $E(x_\alpha) = 0$, in particular, we have

$$E\{\exp(i\sum z_\nu x_{\alpha_\nu})\} = \exp\left\{-\frac{1}{2}\|\sum z_\nu x_{\alpha_\nu}\|^2\right\}$$

$$\left(\|x\|^2 = \int |x(\omega)|^2 P(d\omega)\right).$$

We generalize this property to a family of complex random variables x_α, $\alpha \in A$, and say that it is a complex normal system if for arbitrary α_ν and $z_\nu \in \boldsymbol{C}$ $(= R^1 + iR^1)$,

$$E\{\exp(iRe(\sum \overline{z}_\nu x_{\alpha_\nu}))\} = \exp\left\{-\frac{1}{4}\|\sum \overline{z}_\nu x_{\alpha_\nu}\|^2\right\} \tag{3.6.1}$$

holds. Here, Re refers to the real part of a complex number, and $\overline{z}$ denotes the complex conjugate of z. In particular, we have

$$E\{\exp iRe(\overline{z}x_\alpha)\} = \exp\left\{-\frac{|z|^2}{4}\|x_\alpha\|^2\right\}.$$

If we denote by x'_α, x''_α the real and imaginary parts of x_α, respectively, and let $z = z' + iz''$ (z', z'' are the real and imaginary parts of z, respectively), then we obtain from the identity above

$$E\{\exp\{i(z'x'_\alpha + z''x''_\alpha)\}\} = \exp\left\{-\frac{\|x_\alpha\|^2}{4}(z'^2 + z''^2)\right\},$$

which shows that x'_α and x''_α are independent random variables and both have the same 1-dimensional normal distribution $N(\bullet; 0, \frac{\|x_\alpha\|^2}{2})$.

THEOREM 3.6.1. *Let x_α, $\alpha \in A$, be a system of complex random variables for which $E(x_\alpha) = 0$. If we let*

$$v(\alpha, \beta) = E(x_\alpha \overline{x}_\beta) = (x_\alpha, x_\beta),$$

then $(v(\alpha, \beta))$ is positive definite; namely,

$$\sum_{i,j} \overline{\xi}_i \xi_j v(\alpha_i, \alpha_j) \geq 0. \tag{3.6.2}$$

Conversely, if $(v(\alpha, \beta))$ is positive definite, then there exists a complex normal system x_α, $\alpha \in A$, for which (3.6.1) is valid.

PROOF OF THE FIRST PART.

$$\sum_{i,j} \overline{\xi}_i \xi_j v(\alpha_i, \alpha_j) = \left\|\sum_i \overline{\xi}_i x_{\alpha_i}\right\|^2 \geq 0. \qquad \square$$

PROOF OF THE SECOND PART. Let $\Omega = \boldsymbol{C}^A = R^{2A}$, and introduce a (real-valued) normal distribution N on $\Omega(\boldsymbol{B}^{2A})$ in the following way. Let $\boldsymbol{C}_0^A$ be the set of all the points in $\boldsymbol{C}^A$ with all but a finite number of coordinates being equal to 0. (This set coincides with the subset R_0^{2A} of R^{2A} introduced in §1.4.) For $\boldsymbol{z} = \prod_\alpha z_\alpha \in \boldsymbol{C}_0^A$, set

$$\varphi(\boldsymbol{z}) = \exp\left\{-\frac{1}{4}\sum_{\alpha,\beta} \overline{z}_\alpha z_\beta v(\alpha, \beta)\right\}.$$

Since $\boldsymbol{z} \in \boldsymbol{C}_0^A$, the sum $\sum_{\alpha,\beta}$ above is a finite sum, and hence there is no problem about the convergence. In order to show that this $\varphi(\boldsymbol{z})$ is a characteristic function of some normal distribution N defined over $R^{2A}(\boldsymbol{B}^{2A})$, it is enough to show that $\sum_{\alpha\beta} \overline{z}_\alpha z_\beta v(\alpha,\beta)$ is a positive definite real-valued quadratic form as a function of $\boldsymbol{z} \in R_0^{2A}$. But this fact follows from the assumption that $(v(\alpha,\beta))$ is positive definite. In fact, from the assumption of positive definiteness we get that $v(\alpha,\beta) = \overline{v(\beta,\alpha)}$, and this in turn implies that $\sum_{\alpha,\beta}\cdots$ is a real quadratic form of $\boldsymbol{z} \in R^{2A}$ and in view of (3.6.2) it is positive definite as well. We let $\Omega(\boldsymbol{B}^{2A}, N)$ be the underlying probability space, and for $\omega \in \Omega = \boldsymbol{C}^A$ define $x_\alpha(\omega)$ to be the α-th coordinate (which is a complex number) of ω. Then we have for $\boldsymbol{z} = \prod_\alpha z_\alpha \in \boldsymbol{C}_0^A$

$$
\begin{aligned}
E\{\exp(iRe\sum_\alpha \overline{z}_\alpha x_\alpha)\} &= E\{\exp(i\sum_\alpha (z'_\alpha x'_\alpha + z''_\alpha x''_\alpha))\} \\
&= \exp\left\{-\frac{1}{4}\sum_{\alpha,\beta} \overline{z}_\alpha z_\beta v(\alpha,\beta)\right\}
\end{aligned}
$$
$$(z_\alpha = z'_\alpha + iz''_\alpha,\ x_\alpha = x'_\alpha + ix''_\alpha).$$

If we substitute $t \cdot z_\alpha$ for z_α in the equation above, we get

$$E\{\exp(itRe\sum_\alpha \overline{z}_\alpha x_\alpha)\} = \exp\left\{-\frac{t^2}{4}\sum_{\alpha,\beta} z_\alpha \overline{z}_\beta v(\alpha,\beta)\right\}.$$

Differentiating both sides of the last equation twice with respect to t, and putting $t = 0$, we obtain

$$
\begin{aligned}
&\frac{1}{4}\sum_\alpha \overline{z}_\alpha^2 E(x_\alpha^2) + \frac{1}{4}\sum_\alpha z_\alpha^2 E(\overline{x}_\alpha^2) + \frac{1}{2}\sum_{\alpha,\beta} \overline{z}_\alpha z_\beta (x_\alpha, x_\beta) \\
&\qquad = \sum_{\alpha,\beta} \overline{z}_\alpha z_\beta v(\alpha,\beta),
\end{aligned}
$$

from which it follows that

$$E(x_\alpha^2) = 0,\ (x_\alpha, x_\beta) = v(\alpha,\beta),$$

and thus we can conclude that

$$E\{\exp(iRe\sum_\alpha \overline{z}_\alpha x_\alpha)\} = \exp\left\{-\frac{1}{4}\left\|\sum_\alpha \overline{z}_\alpha x_\alpha\right\|^2\right\}. \qquad \square$$

REMARK. We obtained the fact $E(x_\alpha^2) = 0$ above as a by-product of the calculation, but this is an interesting property of the complex normal process. Indeed, if we write $x_\alpha = x'_\alpha + ix''_\alpha$, then as we noted before, x'_α and x''_α are independent, and they both have the same distribution $N(\bullet; 0, \|x_\alpha\|^2/2)$, and therefore, we get

$$E(x_\alpha^2) = E(x'^2_\alpha) - E(x''^2_\alpha) + 2iE(x'_\alpha)E(x''_\alpha) = 0.$$

THEOREM 3.6.2. *If x_α, $\alpha \in A$, is a complex normal system, which is pairwise orthogonal, then it is an independent system.*

PROOF.

$$
\begin{aligned}
E\{\exp(iRe\sum_\alpha \bar{z}_\alpha x_\alpha)\} &= \exp\left\{-\frac{1}{4}\left\|\sum_\alpha \bar{z}_\alpha x_\alpha\right\|^2\right\}, \\
&= \exp\left\{-\frac{1}{4}\sum_\alpha |z_\alpha|^2\|x_\alpha\|^2\right\} \quad \text{(due to the orthogonality)} \\
&= \prod_\alpha \exp\left\{-\frac{1}{4}|z_\alpha|^2\|x_\alpha\|^2\right\} \\
&= \prod_\alpha E\{\exp(iRe\bar{z}_\alpha x_\alpha)\},
\end{aligned}
$$

from which it follows that the distribution of $\boldsymbol{x} = \prod_\alpha x_\alpha$ is the direct product of the distributions of x_α, $\alpha \in A$. □

THEOREM 3.6.3. *If x_α, $\alpha \in A$, is a complex normal system, then both $Re(x_\alpha)$, $\alpha \in A$, and $Im(x_\alpha)$, $\alpha \in A$ ($Im(z)$ refers to the imaginary part of the complex number z), are real normal systems.*

PROOF. The assertions of the theorem follow immediately by setting $Im(z_\nu) = 0$ or $Re(z_\nu) = 0$, respectively, in (3.6.1). □

The following two theorems also can be deduced immediately from the definition of the complex normal system.

THEOREM 3.6.4. *If x_α, $\alpha \in A$, and y_α, $\alpha \in A$, are both real normal systems such that they are independent and identically distributed, then $x_\alpha + iy_\alpha$, $\alpha \in A$, is a complex normal system.*

THEOREM 3.6.5. *If x_α, $\alpha \in A$, is a complex normal system, and for each $\beta \in B$, y_β is a complex linear combination of elements of x_α, $\alpha \in A$, or the norm limit of a sequence of such linear combinations, then y_β, $\beta \in B$, is also a complex normal system.*

EXAMPLE 3.6.1. **Complex Wiener Process**. If x_t, $-\infty < t < \infty$, is a complex normal system satisfying the property of orthogonal increments, namely, for $s < t \le u < v$,

$$(x_t - x_s, x_v - x_u) = 0,$$

then x_t, $-\infty < t < \infty$, is called a complex Wiener process. For such a process and for $s_1 < t_1 \le s_2 < t_2 \le \cdots \le s_n < t_n$, $x_{t_i} - x_{s_i}$, $i = 1, 2, \cdots, n$, is a complex normal system in view of Theorem 3.6.5, and since the increments are pairwise orthogonal, they are independent by Theorem 3.6.2. Therefore, x_t, $-\infty < t < \infty$, is a complex additive process. The distribution of $x_t - x_s$ is invariant under rotations around the origin of the complex plane. When $s < t < u$, $(x_u - x_t, x_t - x_s) = 0$ holds and therefore, we have

$$\|x_u - x_s\|^2 = \|x_u - x_t\|^2 + \|x_t - x_s\|^2.$$

Consequently, there exists an increasing function $F(t)$ determined uniquely up to an additive constant for which

$$\|x_t - x_s\|^2 = F(t) - F(s)$$

is satisfied. If x_t is right continuous in t in the sense of norm, then $F(t)$ becomes a right continuous function.

Conversely, let us show that for an increasing function $F(t)$ there exists a complex Wiener process x_t for which $\|x_t - x_s\|^2 = F(t) - F(s)$ holds. Let us denote by A the family of all intervals $(a, b]$, $-\infty < a < b < \infty$, and for $\alpha, \beta \in A$, let

$$v(\alpha, \beta) = \begin{cases} F(d) - F(c), & \text{if } \alpha \cap \beta \ (\neq \emptyset) = (c, d], \\ 0, & \text{if } \alpha \cap \beta = \emptyset. \end{cases}$$

Then $(v(\alpha, \beta))$ is positive definite. In fact, if we denote by $c(t, \alpha)$ the indicator function of the interval α, then

$$\begin{aligned} \sum_{i,j} \bar{\xi}_i \xi_j v(\alpha_i, \alpha_j) &= \sum_{i,j} \bar{\xi}_i \xi_j \int c(t, \alpha_i) c(t, \alpha_j) dF(t) \text{ (Riemann-Stieltjes Integral)} \\ &= \int \left| \sum_i c(t, \alpha_i) \bar{\xi}_i \right|^2 dF(t) \geq 0. \end{aligned}$$

Therefore, there exists a complex normal system $\{x_\alpha\}$ satisfying the identity

$$(x_\alpha, x_\beta) = v(\alpha, \beta).$$

If $s < t < u$, then

$$\begin{aligned} \|x_{(s,t]} + x_{(t,u]} - x_{(s,u]}\|^2 &= F(t) - F(s) + F(u) - F(t) \\ &\quad + F(u) - F(s) - 2(F(t) - F(s)) - 2(F(u) - F(t)) = 0, \end{aligned}$$

which implies that

$$x_{(s,t]} + x_{(t,u]} = x_{(s,u]} \text{ holds a.e.}$$

Therefore, if we set for $a < \min(0, t)$

$$x_t = x_{(a,t]} - x_{(a,0]},$$

then x_t is determined except on a set of probability 0 independently of the choice of a, and x_t, $-\infty < t < \infty$, thus obtained gives the desired complex Wiener process.

In case $F(t) = t$, then the process obtained is temporally homogeneous, and this is the process N. Wiener treated as the **Brownian motion**.

3.7. Normal Stationary Processes

If a complex normal system x_t, $-\infty < t < \infty$, is weakly stationary, we call it a **complex normal weakly stationary process**. We have the following theorem corresponding to Theorem 3.1.2.

THEOREM 3.7.1. *A complex normal weakly stationary process is strongly stationary.*

PROOF. Let x_t, $-\infty < t < \infty$, be a complex normal weakly stationary process. Then for an arbitrary choice of $t_1 < t_2 < \cdots < t_n$ and h, we have

$$\begin{aligned} E\{\exp(iRe(\sum \bar{z}_\nu x_{t_\nu}))\} &= \exp\left\{-\frac{1}{4}\sum_{\mu,\nu} \bar{z}_\mu z_\nu (x_{t_\mu}, x_{t_\nu})\right\} \\ &= \exp\left\{-\frac{1}{4}\sum_{\mu,\nu} \bar{z}_\mu z_\nu v(t_\mu - t_\nu)\right\} \quad ((x_t, x_s) = v(t-s)) \\ &= E\{\exp(iRe(\sum \bar{z}_\nu x_{t_\nu + h}))\}, \end{aligned}$$

which implies that the distribution of (x_{t_ν}) and that of $(x_{t_\nu+h})$ coincide and hence (x_t) is strongly stationary. □

According to this theorem, we see that we do not have to distinguish weak and strong stationarity in case of complex normal processes, and we can call them **complex normal stationary processes**. In fact, since we have been assuming that stationary processes are complex valued, it is enough to call them simply **normal stationary processes**.

We proved before Khinchin's Theorem stating that $v(t) = (x_{s+t}, x_s)$ admits the spectral decomposition. However, we have not dealt with the question of whether for such a function $v(t)$ there exists a stationary process (x_t) which satisfies $(x_t, x_s) = v(t-s)$. In fact, such a stationary process exists. We can show even that there exists a normal stationary process satisfying this condition for a given $v(t)$.

THEOREM 3.7.2. *Suppose $v(t)$ is given by the integral $\int e^{-i2\pi t\lambda} dF(\lambda)$, where $F(\lambda)$ is a bounded, right continuous, increasing function. Then there exists a normal stationary process x_t, $-\infty < t < \infty$, for which*

$$(x_t, x_s) = v(t-s)$$

is valid.

PROOF. For arbitrary t_μ, ξ_μ, we have

$$\sum_{\mu,\nu} \bar{\xi}_\mu \xi_\nu v(t_\mu - t_\nu) = \int |\sum_\mu \bar{\xi}_\mu e^{-i2\pi t_\mu \lambda}|^2 dF(\lambda) \geq 0.$$

Therefore, from Theorem 3.6.1 of the preceding section follows immediately the existence of the desired (x_t). □

Next, let us consider the spectral decomposition of x_t due to Kolmogorov:

$$x_t = \int e^{-i2\pi t\lambda} dy_\lambda.$$

As we have shown in the proof of Kolmogorov's spectral decomposition, y_λ belongs to the Hilbert space $\boldsymbol{H}_0$ spanned by the x_t's, and therefore, by Theorem 3.6.5 in the preceding section, we see that y_λ, $-\infty < \lambda < \infty$, is also a complex normal system. Furthermore, since y_λ, $-\infty < \lambda < \infty$, possesses orthogonal increments, it is a complex Wiener process. We can write by using the spectral function F of Khinchin

$$\|y_\lambda - y_\mu\|^2 = F(\lambda) - F(\mu) \quad (\lambda > \mu).$$

Using Theorem 3.4.2, we see that the following holds.

THEOREM 3.7.3. *A normal stationary process is the Fourier transform of the derivative (in Schwartz' distribution sense) of a complex Wiener process.*

3.8. Wiener Integrals and Multiple Wiener Integrals

Let x_t, $-\infty < t < \infty$, be a complex Wiener process and let

$$\|x_t - x_s\|^2 = F(t) - F(s) \quad (t > s).$$

Since x_t possesses orthogonal increments, we can define an integral of the form

$$I(f) = \int f(t)dx_t, \quad f \in L^2(R^1, dF),$$

as we explained in §3.2. For the case of complex Wiener processes (x_t), we call this integral the **Wiener integral**. In this case, we can also define integrals of the following type, which should be called **multiple Wiener integrals**:

$$I_{p,q}(f) = \int \cdots \int f(t_1, \cdots, t_p, s_1, \cdots, s_q)dx_{t_1} \cdots dx_{t_p} d\overline{x}_{s_1} \cdots d\overline{x}_{s_q}.$$

However, in order to define such multiple integrals, we need to assume the continuity of $F(t)$. We will skip detailed discussions of the existence of these multiple integrals in general, but we will explain essential points of the argument for the special case $p = 2, q = 0$:

$$I(f) = I_{2,0}(f) = \iint f(t_1, t_2)dx_{t_1} dx_{t_2}.$$

First, we define $I(f)$ for the indicator function f of a 2-dimensional interval $(a, b] \times (c, d]$ which does not intersect the diagonal line $\{(u, u)/u \in R^1\}$ in R^2, by setting

$$I(f) = (x_b - x_a)(x_d - x_c).$$

The condition that the 2-dimensional interval does not intersect the diagonal line is important. We shall call this condition **condition A**. Next, let us denote by S the set of all linear combinations of the indicator functions of intervals satisfying condition **A**, and for $f \in S$ we extend the definition of $I(f)$ by linearity. It is clear that I becomes a linear operator from S to $\boldsymbol{H} = L^2(\Omega)$. By utilizing condition **A** we can show that

$$\|I(f)\|^2 \leq \|f\|^2 \equiv \iint |f(t_1, t_2)|^2 dF(t_1)dF(t_2)$$

holds. If, in particular, $f(t_1, t_2)$ is a symmetric function of (t_1, t_2), then equality holds in the inequality above, and if f and g are both symmetric, the identity

$$(I(f), I(g)) = (f, g)$$

holds. By using the assumption of continuity of F, we can show that $\overline{S} = L^2 \equiv L^2(R^2, (dF)^2)$, and hence we can extend the definition of $I(f)$ to all of L^2.

The definition of general $I_{p,q}$ can be carried out in a similar manner. Let us denote by $\boldsymbol{H}_{p,q}$ the image of L^2 under the map $I_{p,q}$. In particular, $\boldsymbol{H}_{0,0}$ will denote the 1-dimensional subspace of $L^2(\Omega)$ consisting of all the constant functions. Because of condition **A**, $\{\boldsymbol{H}_{p,q}\}_{p,q}$ gives the pairwise orthogonal family of subspaces of $\boldsymbol{H}$. Next, let us denote by $\boldsymbol{H}^*$ the set of all complex-valued random variables x, which are measurable with respect to x_t, $-\infty < t < \infty$, and satisfy the property

$E(|x|^2) < \infty$. Then we can show that $\boldsymbol{H}^*$ is the direct sum of $\{\boldsymbol{H}_{p,q}\}_{p,q}$. Therefore, it is possible to obtain an orthogonal expansion

$$x = \sum_{p,q} I_{p,q}(f_{p,q})$$

for $x \in \boldsymbol{H}^*$. As functions $f_{p,q}(t_1, \cdots, t_p, s_1, \cdots, s_q)$ in the expansion above, we can choose those which are symmetric both in $(t_1, \cdots, t_p)$ and $(s_1, \cdots, s_q)$, and furthermore, for each pair (p, q), $f_{p,q}$ which is symmetric in this sense can be determined uniquely by x.

EXAMPLE 3.8.1. Let x_t, $-\infty < t < \infty$, be a temporally homogeneous complex Wiener process, and let

$$y_t = \int f(t+s)dx_s, \quad f \in L^2(R^1).$$

Then y_t becomes a normal stationary process. In fact,

$$(y_t, y_u) = \int f(t+s)\overline{f(u+s)}ds = \int f((t-u+s)\overline{f(s)}ds$$

shows that (y_t, y_u) is a function of $t-u$, and hence y_t is stationary. The normality of y_t is obvious since it is a limit of linear combinations of x_t's. If we denote by $\hat{f}$ the inverse Fourier transform of f, then we get

$$(y_t, y_u) = \int e^{-2\pi(t-u)\lambda}|\hat{f}(\lambda)|^2 d\lambda,$$

which shows that $|\hat{f}(\lambda)|^2$ is the density of the spectral function.

EXAMPLE 3.8.2. If we define z_t, $-\infty < t < \infty$, in a similar manner as in the previous example by using a multiple Wiener integral

$$z_t = \iint f_2(t_1+t, t_2+t)dx_{t_1}dx_{t_2} + \int f_1(t_1+t)dx_{t_1},$$

then we obtain a strongly stationary process, but unless $f_2 \equiv 0$, z_t is not normal. We can use even higher-dimensional multiple integrals and limits of such to obtain further examples of strongly stationary processes. It would be an interesting problem to investigate whether a more general strongly stationary process can be constructed by such a method.

3.9. Ergodicity of Normal Stationary Processes

First, let us define the ergodicity and strong mixing property of a strongly stationary process x_t, $-\infty < t < \infty$. If x is a random variable which is measurable with respect to x_t, $-\infty < t < \infty$, then there exists a sequence of Baire functions $\{f_n\}$ of n complex variables for each n and a sequence $\{t_1, t_2, \cdots\}$ of real numbers such that

(3.9.1) $$f_n(x_{t_1}, \cdots, x_{t_n}) \to x \quad \text{(convergence in probability)}$$

holds. Since x_t is strongly stationary, we have

$$\begin{aligned} &P\{\omega/|f_n(x_{t_1+t}, \cdots, x_{t_n+t}) - f_m(x_{t_1+t}, \cdots, x_{t_m+t})| > \epsilon\} \\ &= P\{\omega/|f_n(x_{t_1}, \cdots, x_{t_n}) - f_m(x_{t_1}, \cdots, x_{t_m})| > \epsilon\} \\ &\to 0 \ (m, n \to \infty), \end{aligned}$$

and therefore, the sequence $f_n(x_{t_1+t}, \cdots, x_{t_n+t})$, $n = 1, 2, \cdots$, converges in probability to some random variable x'. x' is determined with probability 1 depending only on x and t, and is independent of the choice of sequences $f_n(x_{t_1}, \cdots, x_{t_n})$ in (3.9.1), and thus we can write $x' = T_t x$. It is clear that

$$T_{t+s}x = T_t T_s x \text{ (a.e.)} \tag{3.9.2}$$

holds. If $T_t x = x$ holds for all t, x is called an **invariant** random variable. Clearly, $x \equiv$ constant is an invariant random variable, but if there exist no other invariant random variables, then we say that the process x_t, $-\infty < t < \infty$, is **ergodic**. If x is invariant, then for any $M > 0$, the truncation $x^{(M)}$ of x defined by

$$x^{(M)}(\omega) = \begin{cases} x(\omega), & |x(\omega)| \le M, \\ 0, & |x(\omega)| > M, \end{cases}$$

is also invariant. If x is not a constant, then by taking M sufficiently large we can make $x^{(M)}$ non-constant. Therefore, if there is no bounded non-constant invariant random variable, then x_t, $-\infty < t < \infty$, is ergodic. Next, if, for an arbitrary pair of random variables x, y which are L^2-norm bounded and are measurable with respect to x_t, $-\infty < t < \infty$,

$$E(T_t x \cdot \overline{y}) \to E(x)E(\overline{y}), \text{ i.e., } (T_t x, y) \to (x, 1)(1, y),$$

holds, then x_t, $-\infty < t < \infty$, is said to be **strongly mixing**. The strong mixing property is stronger than ergodicity. For, if $T_t x = x$, then from the strong mixing property it follows that

$$E(x^2) = E(T_t x \cdot x) \to E(x)^2,$$

and therefore, $V(x) = E(x^2) - E(x)^2 = 0$, which implies that $x =$ const.

With these preparatory materials on hand we can prove the following:

THEOREM 3.9.1 (G. Maruyama). *In order for a normal stationary process to be ergodic it is necessary and sufficient that the spectral function $F(\lambda)$ be continuous.*

THEOREM 3.9.2. *In order for a normal stationary process to be strongly mixing it is necessary and sufficient that $v(t) \to 0$ $(t \to \infty)$ hold.*

PROOF OF THE NECESSITY PART FOR THEOREM 3.9.1. Let x_t, $-\infty < t < \infty$, be a normal stationary process, and suppose that the spectral function $F(\lambda)$ has a jump discontinuity at $\lambda = \mu$. Let

$$x_t = \int e^{-i2\pi\lambda t} dy_\lambda$$

be the Kolmogorov spectral decomposition of x_t, and let us use the same notations such as U_t, $E(\lambda)$, etc. that we used in the proof of Kolmogorov's Spectral Decomposition Theorem in §3.3. Then we see that $y_\lambda = E(\lambda) \cdot x_0$ also has the jump discontinuity at $\lambda = \mu$, and the magnitude of the jump $z = y_{\mu+0} - y_{\mu-0}$ satisfies the property $U_t z = e^{-i2\pi\mu t} z$. Since T_t is an extension of U_t, we have $T_t z = e^{-i2\pi\mu t} z$. From the definition of T_t it follows that $|T_t z| = T_t |z|$, and therefore, $|z|$ is invariant. Furthermore, since the distribution of z is a rotation invariant normal distribution on the complex plane, and since $\|z\|^2 = F(\mu+0) - F(\mu-0)$ which is > 0 by assumption, this distribution is not a δ-distribution, and therefore, $|z|$ is not a constant. Thus x_t is not ergodic.

PROOF OF THE SUFFICIENCY PART OF THEOREM 3.9.1. Let x be a bounded invariant random variable. As the set of all random variables measurable with respect to y_λ, $-\infty < \lambda < \infty$, coincides with the set of those measurable with respect to x_t, $-\infty < t < \infty$, we see that x is measurable with respect to y_λ, $-\infty < \lambda < \infty$. Then, since $\|x\|^2 < \infty$, it follows from the result obtained in the preceding section that x can be represented in the form

$$(3.9.3)\qquad \begin{aligned} x = \text{const.} + \sum_{p+q>0} \\ \int \cdots \int f_{pq}(\lambda_1, \cdots, \lambda_p, \mu_1, \cdots, \mu_q) dy_{\lambda_1} \cdots dy_{\lambda_p} d\overline{y}_{\mu_1} \cdots d\overline{y}_{\mu_q}. \end{aligned}$$

We assume that functions f_{pq} are symmetric with respect to (λ_i) and also with respect to (μ_j). In order to determine $T_t x$, we can replace $dy_{\lambda_1} \cdots$ by $T_t dy_{\lambda_1} \cdots$ in the formula above. But since $T_t dy_\lambda = e^{-i2\pi\lambda t} dy_\lambda, T_t d\overline{y}_\mu = e^{i2\pi\mu t} d\overline{y}_\mu$ (it should be obvious what is meant by this symbolic statement, although we need a more detailed argument for a more precise statement), we have

$$(3.9.4)\qquad \begin{aligned} T_t x = \text{const.} + \sum_{p+q>0} \int \cdots \int f_{pq}(\lambda_1, \cdots, \lambda_p, \mu_1, \cdots, \mu_q) \\ \times e^{-i2\pi t(\sum \lambda_i - \sum \mu_j)} dy_{\lambda_1} \cdots dy_{\lambda_p} d\overline{y}_{\mu_1} \cdots d\overline{y}_{\mu_q}. \end{aligned}$$

Since the integrand of this equation is also symmetric with respect to (λ_i) and (μ_j), we conclude from the uniqueness of the representation and from the fact that $T_t x = x$ that

$$f_{pq}(\lambda_1, \cdots, \lambda_p, \mu_1, \cdots, \mu_q) e^{-i2\pi t(\sum \lambda_i - \sum \mu_j)} = f_{pq}(\lambda_1, \cdots, \lambda_p, \mu_1, \cdots, \mu_q)$$

holds. Therefore, we must have $f_{pq} = 0$ off the hyperplane $\mathbf{\Pi} : \sum \lambda_i - \sum \mu_j = 0$. On the other hand, because of the continuity of F, we have

$$\int_{\mathbf{\Pi}} dF(\lambda_1) \cdots dF(\lambda_p) dF(\mu_1) \cdots dF(\mu_q) = 0,$$

and therefore, we have $f_{pq} = 0$ (a.e.). Thus, we conclude that $x = $ const., showing the ergodicity of x_t. □

PROOF OF THEOREM 3.9.2. If x_t is strongly mixing, then $v(t) = (x_t, x_0) = (T_t x_0, x_0) \to (x_0, 1)^2 = 0$. So, the necessity of the condition is obvious. Conversely, suppose that $v(t) \to 0$. Then, it follows that $F(\lambda)$ must be continuous. Therefore, random variables x, y which are measurable with respect to x_t, $-\infty < t < \infty$, and norm-bounded can be represented in the form (3.9.3), and by using the orthogonality of $\boldsymbol{H}_{pq}$, explained in the preceding section, we can write

$$(3.9.5)\qquad \begin{aligned} (T_t x, y) =& (x, 1)(1, y) + \sum_{p+q>0} \int \cdots \int f_{pq}(\lambda_1, \cdots, \mu_q) \overline{g}_{pq}(\lambda_1, \cdots, \mu_q) \\ & \times e^{-i2\pi t(\sum \lambda_i - \sum \mu_j)} dF(\lambda_1) \cdots dF(\lambda_p) dF(\mu_1) \cdots dF(\mu_q). \end{aligned}$$

Using the hypothesis that $v(t) \to 0$, we can show that for any pair $-\infty < a < b < \infty$

$$\left| \int_a^b e^{-i2\pi t\lambda} dF(\lambda) \right| \to 0 \ \ (t \to \infty)$$

holds.[†] Hence by approximating f_{pq}, g_{pq} by linear combinations of indicator functions of finite intervals in R^1, we can conclude that each summand in the right-hand side of (3.9.5) tends to 0 as $t \to \infty$, and therefore, if the expansions in (3.9.3) for x and y consist of a finite number of terms, then we get $(T_t x, y) \to (x, 1)(1, y)$ as $t \to \infty$. Finally, for general x, y, we can utilize the fact that $\|T_t x\| = \|x\|$ (and similarly for y) to approximate them by those which have expansions with a finite number of terms to obtain the desired conclusion. □

EXAMPLE 3.9.1. A normal process x_t is strongly stationary, and satisfies the condition $E(|x_0|) < \infty$, and therefore the sample mean

$$x^* = \lim_{A \to \infty} \frac{1}{2A} \int_{-A}^{A} x_t dt$$

exists a.e. It is clear that $T_t x^* = x^*$ holds, and hence if $F(\lambda)$ is continuous, then we have $x^* = \text{const.}$ by Theorem 3.9.1, and therefore, $x^* = E(x^*) = E(x_0)$. This gives a method for computing the value of $E(x_0)$ from a sample process. We can state a similar fact for $v(t)$ as well.

3.10. Generalizations of Stationary Processes

First, we consider a **stationary sequence**. It is a misnomer to call it a generalization (it is a special case of a stationary process, to be more precise), but we take it up in this section. Definitions of weak stationarity, strong stationarity, normality, etc. for a random sequence x_n, $n = 1, 2, \cdots$, are the same as for the case of a stationary process. We assume that $E(x_n) = 0$ is satisfied in this discussion. Khinchin's spectral decomposition of $v(n) = (x_{n+m}, x_m)$ is given by

$$v(n) = \int_0^1 e^{-i2\pi\lambda n} dF(\lambda), \tag{3.10.1}$$

where dF is a measure on $R^1/\text{mod}1$. Kolmogorov's spectral decomposition of x_n is given by

$$x_n = \int_0^1 e^{-i2\pi\lambda n} dy_\lambda, \quad \|dy_\lambda\|^2 = dF(\lambda). \tag{3.10.2}$$

It can be seen that properties of stationary processes appear in stationary sequences almost verbatim, or rather in simpler forms in many cases.

Stationary random distributions. A stochastic process $x_t(\omega)$ can be regarded as a family of functions of t parametrized by ω. In a similar manner, by considering a family of Schwartz' distributions parametrized by ω, we can introduce the notion of a **random distribution**. We can define a random distribution in a strong sense as well as in a weak sense. Let us denote by $\mathfrak{D}$ the set of all infinitely differentiable functions with compact support, and let $\mathfrak{D}'$ be the space of all Schwartz' distributions. If a function $x(\phi, \omega)$ defined for $\phi \in \mathfrak{D}$ and $\omega \in \Omega$ belongs to $\mathfrak{D}'$ as a function of ϕ for all (or almost all) ω, and, for each fixed ϕ, is a measurable function of ω, then $x(\phi, \omega)$ is called a **random distribution in a**

[†]Cf. Theorem 19.3 in K. Itô, *Complex multiple Wiener integral*, Japan J. Math. 22 (1952), 63–86 (Kiyosi Itô Selected Papers, eds. D. W. Stroock and S. R. S. Varadhan, Springer-Verlag, 1986, pp. 217–240).

strong sense. On the other hand, if $x(\phi, \omega)$ belongs to $\boldsymbol{H} = L^2(\Omega)$ as a function of ω for each fixed $\phi \in \mathfrak{D}$, and if the map defined by

$$x: \quad \mathfrak{D} \ni \phi \to x(\phi, \bullet) \in \boldsymbol{H}$$

gives a Schwartz' distribution taking values in $\boldsymbol{H}$, namely if $x \in \mathfrak{D}'_{\boldsymbol{H}}$, then $x(\phi, \omega)$ is called a **random distribution in a weak sense**. If a random distribution in a strong sense satisfies the **strong stationarity** defined in the following way:

$$\begin{aligned} &\text{the distribution of } (x(\phi_1), \cdots, x(\phi_n)) \text{ always coincides with that of} \\ &\qquad (x(\phi_1^{(h)}), \cdots, x(\phi_n^{(h)})) \text{ where } \phi^{(h)}(t) = \phi(t+h), \end{aligned} \tag{3.10.3}$$

then it is called a **strongly stationary random distribution**. If $x_t(\omega)$ is a measurable strongly stationary process, for which $E(|x_0|) < \infty$ is satisfied, then if we define for $\phi \in \mathfrak{D}$,

$$x(\phi, \omega) = \int \phi(t) x_t(\omega) dt, \tag{3.10.4}$$

this gives a strongly stationary random distribution. In fact, from $E(|x_0|) < \infty$, it follows that for almost all ω,

$$\int \frac{|x_t(\omega)|}{1+t^2} dt < \infty \tag{3.10.5}$$

holds. (Consider the expectation of this integral.) From this it follows that $x_t(\omega)$, for almost all ω, is a locally integrable function of t, and therefore, (3.10.4) gives a random distribution in a strong sense. One can show easily that it also satisfies the strong stationarity (3.10.3). If for a random distribution in a weak sense $x(\phi, \omega)$, $\phi \in \mathfrak{D}$,

$$m(\phi) \equiv E(x(\phi)), \quad v(\phi, \psi) \equiv E((x(\phi) - m(\phi))(\overline{x(\psi) - m(\psi)})) \tag{3.10.6}$$

are both invariant under parallel translations, namely,

$$m(\phi^{(h)}) = m(\phi), \quad v(\phi^{(h)}, \psi^{(h)}) = v(\phi, \psi), \tag{3.10.7}$$

then $x(\phi, \omega)$ is called a **weakly stationary random distribution**. For example, a weakly stationary process $x_t(\omega)$ can be regarded as a weakly stationary random distribution if we consider it in the following way:

$$x(\phi) = \int \phi(t) x_t dt \quad \text{(integral is taken in the sense of norm convergence).}$$

In fact, we have

$$\begin{aligned} m(\phi) &= \int \phi(t) m(t) dt = m \int \phi(t) dt, \\ v(\phi, \psi) &= \iint \phi(t) \overline{\psi(s)} v(t, s) dt ds = \iint \phi(t) \overline{\psi(s)} v(t-s) dt ds \\ &= \int v(t) \int \phi(t+s) \overline{\psi(s)} ds dt \\ &= \int v(t) \int \phi(t-s) \check{\psi}(s) ds dt, \quad \check{\psi}(s) = \overline{\psi(-s)}, \\ &= \int v(t) (\phi * \check{\psi})(t) dt \\ &= v(\phi * \check{\psi}), \end{aligned}$$

from which the invariance of m and v under parallel translations follows easily. We also see that $v(\phi, \psi)$ can be written as $v(\phi * \check{\psi})$. Actually, the fact that $v(\phi, \psi)$ depends only on $\phi * \check{\psi}$ can be derived by using only the invariance of $v(\phi, \psi)$ under parallel translations, and therefore, we see that the identity

$$v(\phi, \psi) = v(\phi * \check{\psi}) \tag{3.10.8}$$

is valid for a general weakly stationary random distribution. In this identity we can regard v as an element of $\mathfrak{D}'$, and since

$$v(\phi * \check{\phi}) = v(\phi, \phi) = \|x(\phi) - m(\phi)\|^2 \geq 0,$$

we can write, by using the extension of Bochner's theorem to Schwartz' distributions,

$$v = \mathfrak{D}'\text{-}\lim_{A\to\infty} \int_{-A}^{A} e^{-i2\pi\lambda t} dF(\lambda), \tag{3.10.9}$$

where

$$dF(\lambda) \geq 0, \quad \int \frac{dF(\lambda)}{(1+\lambda^2)^k} < \infty. \tag{3.10.10}$$

This result is an extension of the spectral decomposition of A. Khinchin. We can also show that for $x(\phi)$ there exists a process y_λ with orthogonal increments for which

$$x = \mathfrak{D}'_{\boldsymbol{H}}\text{-}\lim_{A\to\infty} \int_{-A}^{A} e^{-i2\pi\lambda t} dy_\lambda, \quad \|dy_\lambda\|^2 = dF(\lambda) \tag{3.10.11}$$

holds. This fact is an extension of Kolmogorov's spectral decomposition.

Now let x_t, $-\infty < t < \infty$, be a temporally homogeneous complex Wiener process introduced in §3.6. Clearly, x_t itself is not a stationary process. But from the temporal homogeneity $\|dx_t\|^2 = dt$, it follows that if we denote by Dx_t the derivative of x_t in the sense of Schwartz' distribution ($\mathfrak{D}'_{\boldsymbol{H}}$), then Dx_t gives a weakly stationary random distribution. (It is clear that the derivative in t of x_t in the ordinary sense does not exist, since $\|dx_t\|^2 = dt$.) In fact, if we consider $m(\phi), v(\phi, \psi)$ for this Dx_t, then since we have

$$Dx_t(\phi) = -x_t(\phi') = -\int x_t \phi'(t)dt = -\int (x_t - x_a)\phi'(t)dt,$$

where a can be chosen arbitrarily, we get

$$m(\phi) = 0,$$

$$\begin{aligned} v(\phi, \psi) &= E\left\{\iint \phi'(t)\overline{\psi'(s)}(x_t - x_a)(\overline{x_s - x_a})dtds\right\} \\ &= \int_a^\infty \int_a^\infty \phi'(t)\overline{\psi'(s)}(x_t - x_a, x_s - x_a)dtds \\ &\quad \text{(by taking } a \text{ suitably so as to have the supports of } \phi, \psi \text{ lie in } (a, \infty)) \\ &= \int_a^\infty \int_a^\infty \phi'(t)\overline{\psi'(s)}(\min(t,s) - a)dtds \\ &= \delta(\phi * \check{\psi}) \quad (\delta \text{ here denotes the Dirac } \delta \text{ function}). \end{aligned}$$

Thus we see that m and v are invariant under parallel translations, and furthermore, if we denote $v(\phi, \psi) = v(\phi * \breve{\psi})$, then we have $v = \delta$. These facts become clearer if we write symbolically in the following way:

$$\left(\frac{dx_t}{dt}, \overline{\frac{dx_s}{ds}}\right) = 0 \quad (t \neq s),$$

$$\left(\frac{dx_t}{dt}, \overline{\frac{dx_t}{dt}}\right) = \frac{\|dx_t\|^2}{dt^2} = \frac{dt}{dt^2} = \frac{1}{dt}.$$

Furthermore, since

$$\delta = \mathfrak{D}'\text{-}\lim_{A\to\infty} \int_{-A}^{A} e^{-i2\pi\lambda t} d\lambda$$

holds, we see that the spectral function for Dx_t is $F(\lambda) \equiv \lambda$. As

$$\int \frac{d\lambda}{1+\lambda^2} < \infty,$$

the index k appearing in the formula (3.10.10) equals 1 in this case.

Vector-valued stationary processes. It is possible to consider a stationary process $\boldsymbol{x}_t$, $-\infty < t < \infty$, whose values are m-dimensional vectors. Let us explain briefly the case of weakly stationary processes. Without loss of generality we may assume that $\boldsymbol{m}(t) \equiv E(\boldsymbol{x}_t) = 0$. $v(t,s)$ becomes the matrix $(v_{ij}(t,s))$ given by

$$v_{ij}(t,s) = E(x_t^i x_s^j), \quad \text{namely } v(t,s) = E(\boldsymbol{x}_t \times \boldsymbol{x}_s). \tag{3.10.12}$$

When $v_{ij}(t,s)$ depends only on $t-s$, then $\boldsymbol{x}_t$ is said to be weakly stationary. In this case we write $v_{ij}(t,s) = v_{ij}(t-s)$. Concerning the spectral decomposition of $v(t) = (v_{ij}(t))$, H. Cramér obtained a theorem which generalizes Khinchin's theorem in the following way. There exists an $m\times m$ matrix-valued function $F(\lambda) = (F_{ij}(\lambda))$, for which

$$v(t) = \int e^{-i2\pi\lambda t} dF(\lambda). \tag{3.10.13}$$

This $F(\lambda)$ is a hermitian matrix satisfying the following properties:

$\lambda \geq \mu \Rightarrow F(\lambda) - F(\mu) \geq 0$ (≥ 0 here means positive definiteness),

$F(\infty)$ is a well-defined matrix, $F(-\infty) = 0$ matrix,

$F(\lambda + 0) = F(\lambda)$.

Indeed, for an arbitrary sequence $\{a_i\}$, $1 \leq i \leq n$, of complex numbers, $v_a(t) = \sum a_i \bar{a}_j v_{ij}(t)$ satisfies

$$\sum_{\mu\nu} \xi_\mu \bar{\xi}_\nu v_a(t_\mu - t_\nu) = \sum_{\mu\nu}\sum_{ij} \xi_\mu \bar{\xi}_\nu a_i \bar{a}_j (x_{t_\mu}^i, x_{t_\nu}^j) = \|\sum \xi_\mu a_i x_{t_\nu}^i\|^2 \geq 0,$$

and therefore, we can represent $v_a(t)$ as

$$v_a(t) = \int e^{-i2\pi\lambda t} dF_a(\lambda).$$

$\Delta F_a(\lambda)$ is a hermitian quadratic form in $\{a_i\}$, and $\Delta F_a(\lambda) \geq 0$. Therefore, functions $F_{ij}(\lambda)$ can be determined so that $\Delta F_a(\lambda) = \sum a_i \bar{a}_j \Delta F_{ij}(\lambda)$ and $\Delta F(\lambda) = (\Delta F_{ij}(\lambda))$ is positive definite.

An extension of Kolmogorov's spectral decomposition to the vector-valued process can be obtained also.

Stochastic processes depending on time and position. While stochastic processes are usually considered to describe quantities changing randomly with the passage of time, we may also consider random quantities depending on positions as well as on the passage of time. For instance, a state of some randomly changing quantity at time t and at a position $\xi = (\xi_1, \xi_2, \xi_3)$ may be represented by a random variable $x(t, \xi, \omega)$. Let us now write

$$m(t, \xi) = E(x(t, \xi, \omega)),$$
$$v(t, \xi; s, \eta) = E\{(x(t, \xi, \omega) - m(t, \xi))\overline{(x(s, \eta, \omega) - m(s, \eta))}\}.$$

If $m(t, \xi)$ and $v(t, \xi; s, \eta)$ are invariant under parallel translations both in time and in space, then we say that $x(t, \xi, \omega)$ is stationary. In this case, $m(t, \xi) = \text{const.}$ (which we will set equal to 0 in the sequel), and $v(t, \xi; s, \eta)$ becomes a function of $t - s, \xi - \eta = (\xi_1 - \eta_1, \xi_2 - \eta_2, \xi_3 - \eta_3)$, which we denote by $v(t - s, \xi - \eta)$. We can obtain the spectral decomposition for this v in the form

$$v(t, \xi) = \int_{R^4} e^{-i2\pi(\lambda + (\sigma, \xi))} dF(\lambda, \sigma), \quad (\sigma, \xi) = \sigma_1\xi_1 + \sigma_2\xi_2 + \sigma_3\xi_3,$$

which corresponds to the Khinchin spectral decomposition. The spectral decomposition of $x(t, \xi, \omega)$ itself, corresponding to the Kolmogorov spectral decomposition, is also possible in the same way as for stationary processes.

As for positions, we may sometimes assume, in addition to invariance under translations, the isotropic property. In this case, $v(t, \xi)$ can be written as $v(t, |\xi|)$, where $|\xi|$ denotes the length of the vector ξ, and the measure $dF(\lambda, \sigma)$ introduced above also has the isotropic property and can be represented in the form $dF(\lambda, \sigma) = dG(\lambda, r) \cdot d\theta$, where r represents the length of σ, and θ denotes the point on the unit sphere in R^3 where the vector σ intersects. $v(t, |\xi|)$ is given by

$$v(t, |\xi|) = \int_{R^2} e^{-i2\pi\lambda t} K(r \cdot |\xi|) dG(\lambda, r), \tag{3.10.14}$$

where the function K can be represented in the following form by using a Bessel function:

$$K(p) = 2\pi \frac{J_{1/2}(2\pi p)}{p^{1/2}}. \tag{3.10.15}$$

Isotropic turbulence. In the preceding paragraphs we considered the situation in which to the pair t, ξ, where t is a point in time and ξ is a point in space, a random scalar $x(t, \xi, \omega)$ corresponded. In this subsection we consider the case where, instead of scalar quantities, vectorial quantities $\boldsymbol{u}(t, \xi, \omega)$ correspond. For example, this occurs if we set the velocity of turbulence at arbitrary time t and a space point ξ to be $\boldsymbol{u}(t, \xi, \omega)$. In order to emphasize only the essentially difficult points in our discussion, we fix the time t and write the vector-valued function as $\boldsymbol{u}(\xi, \omega)$, and assume that $E(\boldsymbol{u}(\xi, \omega)) = 0$. Then the quantity $v(\xi, \eta)$ is given by

$$v(\xi, \eta) = E[\boldsymbol{u}(\xi, \omega) \otimes \boldsymbol{u}(\eta, \omega)]$$

and this is an element of the tensor product $T_\xi \otimes T_\eta$ of the tangent planes T_ξ, T_η at ξ, η, respectively. If this quantity is invariant under orthogonal transformations in the space R^3, $\boldsymbol{u}(\xi, \omega)$ is called an isotropic turbulence. Let us make the meaning of this invariance more precise. Let g be an arbitrary orthogonal transformation. g moves a point ξ to the point $g \cdot \xi$, and at the same time it induces an orthogonal transformation $\dot{g}$ from the tangent plane T_ξ at the point ξ to the tangent plane $T_{g\cdot\xi}$ at the point $g\cdot\xi$. Furthermore, it induces an orthogonal transformation from $T_\xi \otimes T_\eta$ to $T_{g\cdot\xi}\otimes T_{g\cdot\eta}$ as well. We denote this induced orthogonal transformation by the same symbol $\dot{g}$. The invariance of $v(\xi,\eta)$ under g means

$$E[\dot{g}\boldsymbol{u}(\xi,\omega)\otimes \dot{g}\boldsymbol{u}(\eta,\omega)] = E[\boldsymbol{u}(g\cdot\xi,\omega)\otimes\boldsymbol{u}(g\cdot\eta,\omega)], \tag{3.10.16}$$

namely,

$$\dot{g}v(\xi,\eta) = v(g\xi, g\eta).$$

Suppose we choose an orthogonal system (e_1,e_2,e_3) in the space of ξ, and in the tangent plane T_ξ an orthogonal system $(e_1(\xi),e_2(\xi),e_3(\xi))$ parallel to (e_1,e_2,e_3), and determine coordinates accordingly. Then we have

$$\begin{cases} \xi' = g\xi \iff \xi'_i = \sum_j g_{ij}\xi_j + h_i, \quad (g_{ij}) = \text{ orthogonal matrix}, \\ \qquad\qquad u' = \dot{g}u \iff u'_i = \sum_j g_{ij}u_j \end{cases} \tag{3.10.17}$$

so that (3.10.16) becomes

$$E\left(\sum_k g_{ik}u_k(\xi)\sum_\ell g_{j\ell}u_\ell(\eta)\right) = E(u_i(g\xi)\cdot u_j(g\eta)),$$

namely

$$\sum_{k\ell} g_{ik}g_{j\ell}v_{k\ell}(\xi,\eta) = v_{ij}(g\xi,g\eta). \tag{3.10.18}$$

If (g_{ij}) = identity matrix, in particular, then we have

$$v_{ij}(\xi,\eta) = v_{ij}(\xi+h,\eta+h), \quad h=(h_1,h_2,h_3),$$

which implies that $v_{ij}(\xi,\eta) = v_{ij}(\xi-\eta,0)$ and hence depends only on the difference $\xi-\eta$. We denote it as $v_{ij}(\xi-\eta)$. Then we have from (3.10.18)

$$\sum_{k\ell} g_{ik}g_{j\ell}v_{k\ell}(\xi) = v_{ij}(g\xi).$$

If we define $v(\xi;a,b) = \sum_{ij} a_ib_jv_{ij}(\xi)$, then from the equation above we get

$$v(g\xi; ga, gb) = v(\xi;a,b), \quad \text{where } g \text{ is an orthogonal matrix.} \tag{3.10.19}$$

From the definition we can see that $v(\xi;a,a)$ is a positive definite function of ξ so that we can write

$$v(\xi;a,a) = \int e^{-i2\pi(\lambda,\xi)}m(d\lambda;a), \quad m(d\lambda,a)\geq 0.$$

If we use the fact that $v(\xi,a,b)$ is a bilinear form on a,b, we can deduce that

$$v(\xi;a,b) = \int e^{-i2\pi(\lambda,\xi)}m(d\lambda;a,b), \ \ m(d\lambda;a,a) = m(d\lambda;a)\geq 0, \tag{3.10.20}$$

where $m(d\lambda;a,b)$ is a positive definite bilinear form on a,b. Furthermore, from the invariance of v indicated by (3.10.19), we can show that m is invariant as well:

$$m(gd\lambda;ga,gb) = m(d\lambda;a,b).$$

From this fact, we can deduce that m can be written in the following form:

$$m(d\lambda;a,b) = \sum a_ib_j[\theta_i\theta_jd\theta m_1(dr) + (\delta_{ij}-\theta_i\theta_j)d\theta m_2(dr) + m_0(d\lambda)]. \tag{3.10.21}$$

Here, $r = |\lambda|$, $\theta_i = \lambda_i/|\lambda|$, $d\theta =$ the surface element of the unit sphere of R^3, m_1, m_2 are finite measures on $(0, \infty)$, and m_0 is a point mass supported on the origin of R^3 (λ-space). Conversely, starting with m given in the form (3.10.21) and determining v by means of (3.10.20), we can obtain an isotropic turbulence.

CHAPTER 4

Markov Processes

4.1. Conditional Probability

Let us begin by taking $\Omega(\boldsymbol{B}, P)$ to be our basic probability space. Let $\boldsymbol{B}_1$ be a Borel sub-field of $\boldsymbol{B}$. Sets or functions measurable with respect to $\boldsymbol{B}_1$ are naturally measurable with respect to $\boldsymbol{B}$, but the converse may not hold in general. Next, let A be an event; namely, be measurable with respect to $\boldsymbol{B}$. We define, following Doob, the **conditional probability** $P(A/\boldsymbol{B}_1)$ of A with respect to $\boldsymbol{B}_1$ to be the real-valued function $P(A/\boldsymbol{B}_1)(\omega)$ of ω satisfying the following properties:

(C.1) $P(A/\boldsymbol{B}_1)(\omega)$ is measurable with respect to $\boldsymbol{B}_1$ (and hence, of course, is a random variable on $\Omega(\boldsymbol{B}, P)$).

(C.2) If B is measurable with respect to $\boldsymbol{B}_1$, namely, if $B \in \boldsymbol{B}_1$, then

$$P(A \cap B) = \int_B P(A/\boldsymbol{B}_1)(\omega) P(d\omega). \tag{4.1.1}$$

By using the Radon-Nikodým theorem, we can show that there is such a function $P(A/\boldsymbol{B}_1)(\omega)$ and it is unique up to a set of measure 0. Indeed, if we regard $P(A \cap B)$ as a set function in B, then it is a finite measure on $\Omega(\boldsymbol{B}_1)$, and from the fact that $P(A \cap B) \leq P(B)$, it follows that it is absolutely continuous with respect to P (more precisely, with respect to P restricted to $\boldsymbol{B}_1$). Therefore, there exists a function measurable with respect to $\boldsymbol{B}_1$ satisfying (4.1.1), and it is unique modulo a set of measure 0.

Let us give some explanation for this definition, since there may be people who might wonder what this definition given out of the blue has to do with what we know as "conditional probability" by common sense. Suppose we partition Ω into a disjoint union of a finite or countable number of measurable subsets:

$$\Omega = B_1 \cup B_2 \cup \cdots \quad \text{(disjoint union)}. \tag{4.1.2}$$

Let us denote by $\boldsymbol{B}_1$ the family of all subsets of Ω that can be represented as a union of some of $B_1, B_2, \cdots$ (may be 0 or a finite number or infinitely many of B_j's). It is clear that $\boldsymbol{B}_1$ is a Borel field of subsets of Ω. The function $P(A/\boldsymbol{B}_1)(\omega)$ measurable with respect to $\boldsymbol{B}_1$ in this case is a function taking constant values $a_1, a_2, \cdots$ on $B_1, B_2, \cdots$, respectively. Therefore, if we substitute B_i for B in (4.1.1), we get

$$P(A \cap B_i) = a_i P(B_i),$$

from which it follows that

$$a_i = P(A \cap B_i)/P(B_i).$$

Thus, $P(A/\boldsymbol{B}_1)(\omega)$ is given by

$$P(A/\boldsymbol{B}_1)(\omega) = P(A \cap B_i)/P(B_i), \quad \omega \in B_i. \tag{4.1.3}$$

Usually, the quantity $P(A \cap B)/P(B)$ is called the probability of A under the condition of B, and it is written as $P(A/B)$. If we use this notation, then (4.1.3) takes the following form:

$$P(A/\boldsymbol{B}_1)(\omega) = P(A/B_i), \quad \omega \in B_i.$$

The relationship between the definition by Doob and the usual definition of conditional probability should now be clear.

Let $\boldsymbol{x}(\omega)$ be a random vector with values in $R^\Lambda(\boldsymbol{B}^\Lambda)$, and $\boldsymbol{B}_1$ be $\{\boldsymbol{x}^{-1}(E)/E \in \boldsymbol{B}^\Lambda\}$. Denote coordinates of $\boldsymbol{x}(\omega)$ by $x_\lambda(\omega)$. Then $\boldsymbol{B}_1$ can be identified with the Borel field generated by sets of the form

$$\{\omega/x_\lambda(\omega) \le c\}.$$

In this case the conditional probability $P(A/\boldsymbol{B}_1)(\omega)$ can be written, since it is $\boldsymbol{B}_1$-measurable, in the form $\varphi(\boldsymbol{x}(\omega))$, where φ is a $\boldsymbol{B}^\Lambda$-measurable function on $R^\Lambda(\boldsymbol{B}^\Lambda)$. Therefore, (4.1.1) takes the form

$$P(A \cap \boldsymbol{x}^{-1}(E)) = \int_{\boldsymbol{x}^{-1}(E)} \varphi(\boldsymbol{x}(\omega))P(d\omega).$$

If we denote by $P_{\boldsymbol{x}}$ the distribution of $\boldsymbol{x}$, then since $P_{\boldsymbol{x}} = P \cdot \boldsymbol{x}^{-1}$, we have

$$P(A \cap \boldsymbol{x}^{-1}(E)) = \int_E \varphi(\xi)P_{\boldsymbol{x}}(d\xi).$$

The left-hand side of the above identity, considered as a set function of E, is a measure on $R^\Lambda(\boldsymbol{B}^\Lambda)$, and in view of the inequality $P(A \cap \boldsymbol{x}^{-1}(E)) \le P(\boldsymbol{x}^{-1}(E)) = P_{\boldsymbol{x}}(E)$, it is absolutely continuous with respect to $P_{\boldsymbol{x}}$. Therefore, the function $\varphi(\xi)$ is uniquely determined by the condition specified by the identity above. This $\varphi(\xi)$ is precisely what Kolmogorov defined as the conditional probability $P(A/\boldsymbol{x} = \xi)$. $P(A/\boldsymbol{B}_1)(\omega)$ due to Doob, which we defined above, is a random variable obtained by substituting $\boldsymbol{x}(\omega)$ for ξ in this $\varphi(\xi)$. We denote this in the sequel as $P(A/\boldsymbol{x})$.

EXAMPLE 4.1.1. Suppose that x and y are two real-valued random variables with the joint distribution having the density function $f(\xi, \eta)$. Then the conditional probability of Kolmogoroff is given by

$$P(y \in E/x = \xi) = \int_E f(\xi, \eta)d\eta \Big/ \int_{R^1} f(\xi, \eta)d\eta,$$

from which we can deduce that $P(y \in E/x)$ of Doob can be written as follows:

$$P(y \in E/x) = \int_E f(x, \eta)d\eta \Big/ \int_{R^1} f(x, \eta)d\eta.$$

4.2. Conditional Expectation

Conditional expectation or **conditional mean** can be defined as an exact parallel of the conditional probability. Let $y(\omega)$ be a real- (or complex-) valued random variable, and suppose that

$$E|y| < \infty \tag{4.2.1}$$

is satisfied. We let $\boldsymbol{B}_1$ be, as in the preceding section, the Borel sub-field of $\boldsymbol{B}$. For an arbitrary $B \in \boldsymbol{B}$ we set

$$E(y; B) = \int_B y(\omega)P(d\omega). \tag{4.2.2}$$

Then we define the conditional expectation $E(y/\boldsymbol{B}_1)$ of y with respect to $\boldsymbol{B}_1$ as a function $E(y/\boldsymbol{B}_1)(\omega)$ of ω satisfying the following properties:

(E.1) $E(y/\boldsymbol{B}_1)(\omega)$ is measurable with respect to $\boldsymbol{B}_1$.

(E.2) For $B \in \boldsymbol{B}_1$, $E(y; B) = E(E(y/\boldsymbol{B}_1); B)$ holds.

Existence and uniqueness modulo a set of P-measure 0 of such a function $E(y/\boldsymbol{B}_1)(\omega)$ can be proved, exactly as in the case for the conditional probability, by using the Radon-Nikodým theorem.

If we denote by $\chi_A(\omega)$ the indicator function of the set A, then $E(\chi_A/\boldsymbol{B}_1)$ coincides with $P(A/\boldsymbol{B}_1)$, and therefore, we may regard the conditional probability as a special case of the conditional expectation. Let us next state properties of the conditional expectation. You may derive corresponding properties of the conditional probability from them.

(i) If z is measurable with respect to $\boldsymbol{B}_1$, and if $E(|y|)$, $E(|zy|) < \infty$, then

$$E(zy/\boldsymbol{B}_1) = zE(y/\boldsymbol{B}_1) \tag{4.2.3}$$

holds. For the proof, first note that if z is the indicator function of a set $M \in \boldsymbol{B}_1$, then for $B \in \boldsymbol{B}_1$, we have

$$\begin{aligned} E(zy; B) = E(y; M \cap B) &= \int_{M \cap B} E(y/\boldsymbol{B}_1)(\omega) P(d\omega) \\ &= \int_B z(\omega) E(y/\boldsymbol{B}_1)(\omega) P(d\omega). \end{aligned}$$

Since $z(\omega)E(y/\boldsymbol{B}_1)(\omega)$ is $\boldsymbol{B}_1$-measurable, we get

$$E(zy/\boldsymbol{B}_1)(\omega) = z(\omega)E(y/\boldsymbol{B}_1)(\omega).$$

As both sides of the identity (4.2.3) are linear in z, we can deduce from the special case above that the identity (4.2.3) is satisfied for an arbitrary $\boldsymbol{B}_1$-measurable z.

(ii) If $\boldsymbol{B}_1 \subset \boldsymbol{B}_2$, then

$$E(y/\boldsymbol{B}_1) = E(E(y/\boldsymbol{B}_2)/\boldsymbol{B}_1) \tag{4.2.4}$$

holds. We see that if $B \in \boldsymbol{B}_1$, then $B \in \boldsymbol{B}_2$ holds, and therefore,

$$\begin{aligned} E(y; B) - E\{E(y/\boldsymbol{B}_2); B\} \\ = E\{E(E(y/\boldsymbol{B}_2)/\boldsymbol{B}_1); B\}, \end{aligned}$$

from which (4.2.4) follows.

(iii) If $y = c_1 y_1 + c_2 y_2$, where c_1, c_2 are constants, and $E|y_i| < \infty$ for $i = 1, 2$, then

$$E(y/\boldsymbol{B}_1) = c_1 E(y_1/\boldsymbol{B}_1) + c_2 E(y_2/\boldsymbol{B}_1) \tag{4.2.5}$$

holds.

(iv) If $y_n \to y$ (a.e.) and $|y_n| \le z$ with $E|z| < \infty$ for all n, then

$$E(y_n/\boldsymbol{B}_1) \to E(y/\boldsymbol{B}_1) \text{ (a.e.)} \tag{4.2.6}$$

holds.

REMARK 1. All of the statements above hold modulo sets of P-measure 0.

REMARK 2. When x is an arbitrary random variable, we can define $E(y/x)$ in the same way as we defined $P(A/x)$.

4.3. Martingales

Although a **martingale** is a very important concept in probability theory, we state here only the facts we need concerning it in the subsequent discussion of this chapter.

Let x_t, $t \in T$, where T is a set of real numbers, be a stochastic process, and assume that $E(|x_t|) < \infty$ holds for all $t \in T$. Suppose for each $t \in T$ there exists a Borel field $\boldsymbol{B}_t \subseteq \boldsymbol{B}$ so that

(i) if $s < t$, then $\boldsymbol{B}_s \subset \boldsymbol{B}_t$ holds.

Furthermore, we suppose that

(ii) x_t is $\boldsymbol{B}_t$-measurable for each $t \in T$.

Now, if

(iii) $E\{x_t/\boldsymbol{B}_s\} = x_s$ holds with probability 1, whenever $s < t$, then we say that $\{x_t\}$ is a martingale with respect to $\{\boldsymbol{B}_t\}$,

while if

(iii′) $E\{x_t/\boldsymbol{B}_s\} \geq x_s$ holds with probability 1, whenever $s < t$, then we say that $\{x_t\}$ is a **sub-martingale** with respect to $\{\boldsymbol{B}_t\}$.

Clearly, a martingale is a special case of a sub-martingale.

Let us state some properties of a sub-martingale. First of all we see that

$$E(x_s) \leq E(x_t) \text{ holds, whenever } s < t.$$

This is obvious from condition (iii′). Since $E(x_t)$ is monotone increasing in t, it is continuous except at at most a countable number of points.

THEOREM 4.3.1. *Assume that T is an interval. Suppose that x_t, $t \in T$, is a sub-martingale with respect to $\{\boldsymbol{B}_t, t \in T\}$. If x_t is separable in the sense of* §2.8, *then a sample process of x_t for almost all ω has only the discontinuity points of the first kind.*

We omit the proof of this theorem.

4.4. Transition Probabilities

Let R be a compact Hausdorff space satisfying the second axiom of countability. Then we may assume that the topology of R is given by some metric on R. Let us denote by $\boldsymbol{B}_R$ the smallest Borel field of R containing all open subsets of R.

A function $P(t, x, E)$ of t, x, E, where $t \geq 0$ is a real parameter representing time, $x \in R$, and $E \in \boldsymbol{B}_R$, is called a **transition probability** if it satisfies the following properties:

(T.1) When t, x are fixed, $P(t, x, E)$ is a probability distribution as a function of E.

(T.2) (continuity in x) For a fixed t, the probability measure $P(t, x, E)$ converges vaguely to $P(t, x_0, E)$ as $x \to x_0$. This means that for an arbitrary continuous function f on R,

$$\int_R f(y)P(t, x, dy) \to \int_R f(y)P(t, x_0, dy), \quad x \to x_0,$$

holds.

(T.3) (continuity in t) For a fixed x, $P(t,x,E)$ converges vaguely to $\delta(x,E) = \begin{cases} 1, & x \in E, \\ 0, & x \in E^c. \end{cases}$ Namely, for an arbitrary continuous function f on R,

$$\int_R f(y)P(t,x,dy) \to \int_R f(y)\delta(x,dy) \equiv f(x), \quad t \to 0,$$

holds.

Using (T.2), we can prove that $P(t,x,E)$ is $\boldsymbol{B}_R$-measurable in x when t, E are fixed. For this purpose it is enough to show that for an arbitrary bounded $\boldsymbol{B}_R$-measurable function f, the function $\varphi(x)$ defined by

$$\varphi(x) = \int_R f(y)P(t,x,dy)$$

is $\boldsymbol{B}_R$-measurable. If we denote by $\mathfrak{F}$ the set of all real-valued functions f on R having this property, then by (T.2), $\mathfrak{F}$ contains all continuous functions. Furthermore, using the definition of measurability it is easy to show that $\mathfrak{F}$ is closed under the operation of taking pointwise limits. Therefore, an arbitrary $\boldsymbol{B}_R$-measurable function belongs to $\mathfrak{F}$. With this remark in mind, we now add the following condition:

(T.4) (**The Chapman-Kolmogorov Equation**)

$$P(t+s,x,E) = \int_R P(t,x,dy)P(s,y,E).$$

By the remark made above, it is clear that the integral on the right-hand side above makes sense.

Intuitively speaking, $P(t,x,E)$ represents the probability that an entity which was at x initially moves into E after the lapse of time t. The Chapman-Kolmogorov equation states a condition which is demanded by a natural consideration that for an entity initially at x to move into E after the lapse of time $t+s$ it is necessary that it moves from x to some point y of R after the lapse of time t and then moves into E after time s.

EXAMPLE 4.4.1. Let R be a finite set. Clearly, R with the discrete topology satisfies the conditions stated above. In order to define $P(t,x,E)$, it suffices to determine its value when E is a set $\{y\}$ consisting of a single point. Denote this value by $P(t,x,y)$. Then, by (T.1) we have

$$P(t,x,y) \geq 0, \quad \sum_y P(t,x,y) = 1. \tag{4.4.1}$$

It is clear that (T.2) is satisfied. Condition (T.3) takes the form:

$$\text{as } t \to 0, \quad P(t,x,y) \to \delta(x,y) = \begin{cases} 1, & x = y, \\ 0, & x \neq y. \end{cases} \tag{4.4.2}$$

When t is fixed $(P(t,x,y), x, y \in R)$ can be considered as a finite matrix, which we denote by P_t. (4.4.1) states that the entries of P_t are always non-negative, and the sum of entries on each row equals 1. A matrix with such properties is called a **stochastic matrix**. Condition (4.4.2) means in this case

$$P_t \to I \text{ (identity matrix)}, \quad t \to 0.$$

The Chapman-Kolmogorov equation is represented by using the matrix multiplication as

$$P_{t+s} = P_t \cdot P_s. \tag{4.4.3}$$

EXAMPLE 4.4.2. Let R be the one-point compactification of the reals R^1 and let ∞ be the adjoined point. For $x \in R^1$, $E \subset R^1$, we set

$$P(t,x,E) = \int_E N_t(y-x)dy, \quad N_t(x) = \frac{1}{\sqrt{2\pi t}} e^{-\frac{x^2}{2t}},$$

and also

$$P(t,\infty,E) = \delta(\infty,E).$$

It is obvious that (T.1) is satisfied. Let us investigate the continuity in x. If $x_0 \neq \infty$, then for any continuous function f on R, we have

$$\begin{aligned}\int f(y)N_t(y-x)dy &= \int f(y+x)N_t(y)dy \\ &\to \int f(y+x_0)N_t(y)dy = \int f(y)N_t(y-x_0)dy.\end{aligned}$$

(Note that since f is continuous on the compact space $R^1 \cup \{\infty\}$, it is, of course, bounded.) If $x_0 = \infty$, then for any y, we have $f(y+x) \to f(\infty)$ $(x \to x_0 = \infty)$, and therefore, in the same way as above

$$\int f(y)N_t(y-x)dy \to f(\infty) = \int f(y)\delta(\infty,dy).$$

The continuity in t follows from

$$\int f(y)N_t(y-x)dy = \int f(x+\sqrt{t}y)N_1(dy) \to f(x), \ t \to 0.$$

The Chapman-Kolmogorov equation is satisfied when $x \neq \infty$ because of the property $N_{t+s} = N_t * N_s$ of the normal distribution, and is obvious when $x = \infty$.

EXAMPLE 4.4.3. Let $R = [0,\infty]$ and define $P(t,x,E)$ as follows:

$$\begin{aligned}&P(t,x,E) = \int_E [N_t(y-x) + N_t(y+x)]dy, \ x \in [0,\infty), \ E \subset [0,\infty), \\ &P(t,\infty,E) = \delta(\infty,E).\end{aligned}$$

It is easy to verify that $P(t,x,E)$ defined above satisfies the four properties (T.1) $\sim$ (T.4) mentioned before.

REMARK. If we use (T.4), we can derive from (T.3) the following property:

(T.3′) If x is fixed, $P(t,x,E)$ converges to $P(t_0,x,E)$ vaguely as $t \to t_0$ (cf. §4.6). This is the reason why we called property (T.3) the continuity in t.

4.5. Semi-Groups and Dual Semi-Groups Associated with Transition Probabilities

We use the same notations as in the preceding section. Let us denote by $\boldsymbol{C}$ the space of all continuous functions on R. $\boldsymbol{C}$ is a vector space under the usual linear operations, and furthermore, by introducing the norm given by $\|f\| = \max_{x\in R}|f(x)|$, it becomes a separable Banach space. Define T_t, $t > 0$, by

$$T_t f(x) = \int_R f(y)P(t,x,dy). \tag{4.5.1}$$

Then, by (T.2) we see that $T_t f \in \boldsymbol{C}$ holds, if $f \in \boldsymbol{C}$, and T_t can be regarded as a linear operator on $\boldsymbol{C}$ into itself. Furthermore, by using (T.1) we can show that

$$T_t \geq 0, \text{ i.e., } f \geq 0 \Rightarrow T_t f \geq 0, \tag{4.5.2}$$

$$T_t 1 = 1. \tag{4.5.3}$$

Consequently, we also have

$$\|T_t\| \equiv \sup\{\|T_t f\|, \|f\| \leq 1\} = 1. \tag{4.5.4}$$

According to (T.3) we have

$$T_t f(x) \to f(x), \quad x \in R, \quad t \to 0. \tag{4.5.5}$$

Since the dual space $\boldsymbol{C}^*$ of $\boldsymbol{C}$ is the space of (signed) regular measures on R, (4.5.5) above is equivalent to the condition that

$$T_t \to I \quad \text{(in the weak operator topology)}, \tag{4.5.6}$$

where I is the identity operator. In other words, for arbitrary $f \in \boldsymbol{C}, \mu \in \boldsymbol{C}^*$,

$$(T_t f, \mu) \to (f, \mu).$$

From (T.4) follows the property of semi-group:

$$T_t T_s = T_{t+s}. \tag{4.5.7}$$

In general, if a family of bounded linear operators T_t, $t > 0$, on a separable Banach space E satisfies

$$\|T_t\| \leq 1, \; T_t \to I \text{ (in the weak operator topology)}, \; T_t T_s = T_{t+s}, \tag{4.5.8}$$

then it is called a semi-group of operators on E, or more simply a **semi-group** on E.

Since the family of operators T_t, $t > 0$, on $\boldsymbol{C}$ satisfies (4.5.7), it is a semi-group on $\boldsymbol{C}$. We call this the **semi-group associated with $\boldsymbol{P(t, x, E)}$**.

Next, define for $\mu \in \boldsymbol{C}^*$,

$$T_t^* \mu(E) = \int_R P(t, x, E) \mu(dx). \tag{4.5.9}$$

Then, since $0 \leq P(t, x, E) \leq 1$, we have

$$\|T_t^* \mu\| \leq \|\mu\|, \text{ i.e., } \|T_t^*\| \leq 1, \tag{4.5.10}$$

where $\|\mu\|$ denotes the total variation of μ. Because of the identity

$$(T_t f, \mu) = (f, T_t^* \mu) = \int_R \int_R f(y) P(t, x, dy) \mu(dx)$$

we see that T_t^* is the adjoint operator of T_t, and hence the use of the notation T_t^* is justified. From the Chapman-Kolmogorov equation, we obtain $T_t^* T_s^* = T_{t+s}^*$. Furthermore, we have

$$(f, T_t^* \mu) = (T_t f, \mu) \to (f, \mu), \quad f \in \boldsymbol{C}, \quad \mu \in \boldsymbol{C}^*.$$

Since $\boldsymbol{C}$ does not coincide with $(\boldsymbol{C}^*)^*$, this convergence is not the same as $T_t^* \to I^*$ (in the weak operator topology), where I^* denotes the identity operator on $\boldsymbol{C}^*$, but since it is fairly close, we express this convergence as $T_t^* \to I^*$ (in the weak* sense).

Summarizing the conditions obtained above, we have

$$\|T_t^*\| \leq 1, \; T_t^* \to I^* \text{ (in the weak* sense)}, \; T_t^* T_s^* = T_{t+s}^*. \tag{4.5.11}$$

Since $\boldsymbol{C}^*$ is not separable, and furthermore, since the second condition above differs from the corresponding one in (4.5.7), we cannot exactly call T_t^*, $t > 0$, a semi-group on $\boldsymbol{C}^*$, but since it is very close to such a semi-group, we call it the **dual semi-group** associated with $P(t, x, E)$.

4.6. Hille-Yosida Theory (i)

Let E be a separable Banach space, and let T_t, $t > 0$, be a semi-group on E. Although in (4.5.8) of the preceding section, we required as the second condition for a semi-group that $T_t \to I$ (in the weak operator topology), we can actually derive from this, by using other properties, the following stronger property:

(4.6.1) $$\|T_t f - f\| \to 0, \text{ i.e., } T_t \to I \text{ (in the strong operator topology).}$$

Since we need a theorem due to N. Dunford (see Annals of Math., vol. 33, 1932, pp. 567-573) to prove this fact, we shall omit the proof here. From (4.6.1), we obtain further that

(4.6.2) $$\|T_t f - T_s f\| \to 0 \quad (t \to s).$$

In order to show this, we set $u = \min(t, s)$, $v = |t - s|$ and observe

$$\|T_t f - T_s f\| = \|T_u(T_v f - f)\| \le \|T_v f - f\| \to 0.$$

As can be seen from this proof, the convergence in (4.6.2) is uniform on s, and hence $T_t f$ is uniformly continuous in t.

Hille-Yosida Theory is concerned with the generator of a semi-group. By the semi-group property $T_t T_s = T_{t+s}$, it is not necessary to know T_t for all $t > 0$ but only for $0 < t < \delta$, in order to determine T_t for arbitrary $t > 0$. Here, δ can be made arbitrarily small as long as it remains positive. T_t, $0 < t < \delta$, may be regarded as a **germ** for the semi-group T_t, $t > 0$. As the quantity which determines the limiting behavior as $\delta \to 0$, we consider the operator A given by

(4.6.3) $$Af = \lim_{t \downarrow 0} \frac{T_t f - f}{t} \quad \text{(limit taken in the sense of norm on } E\text{)},$$

and call it the **generator** or the **infinitesimal generator** of T_t. The domain $\mathfrak{D}(A)$ of definition of A is, by definition, the set of all $f \in E$ for which the above limit exists.

Clearly, $\mathfrak{D}(A)$ is a linear subspace of E and A is a linear operator, but in general, $\mathfrak{D}(A)$ does not coincide with E, nor is A bounded. In order to clarify the situation, however, let us at first suppose that A is bounded and $\mathfrak{D}(A) = E$, and carry on formal calculations. We assume furthermore that

(4.6.4) $$A = \lim_{t \downarrow 0} \frac{T_t - I}{t}$$

holds in the sense of the operator norm convergence. Then, from these assumptions, it follows, since

$$\frac{T_{t+\delta} - T_t}{\delta} = \frac{T_\delta - I}{\delta} \cdot T_t,$$

that we obtain by letting $\delta \to 0$,

$$\frac{dT_t}{dt} = AT_t.$$

Solving this equation just as for scalar-valued functions, we get

$$T_t = e^{tA}\left(=\sum_{n=0}^{\infty}\frac{(tA)^n}{n!}\right).$$

If we consider the Laplace transform R_λ of T_t, i.e.,

$$R_\lambda = \int_0^\infty e^{-\lambda t}T_t dt\left(=\lim_{n\to\infty}\frac{1}{n}\sum_{k=0}^{\infty}e^{-\lambda\frac{k}{n}}T_{\frac{k}{n}}\right),$$

then by computing formally, we get

$$R_\lambda = \int_0^\infty e^{-\lambda t}e^{tA}dt = \int_0^\infty e^{-t(\lambda I - A)}dt = (\lambda I - A)^{-1}.$$

Since $\lambda I f = \lambda f$, we may write simply λ for λI so that

$$R_\lambda = (\lambda - A)^{-1}.$$

This means that $u = R_\lambda v$ satisfies the equation

$$(\lambda - A)u = v.$$

In this sense, R_λ is called the resolvent of T_t.

Keeping the formal arguments made above in mind, let us give a more rigorous treatment of the general case. Let us define the resolvent R_λ by

$$(R_\lambda f, \mu) = \int_0^\infty e^{-\lambda t}(T_t f, \mu)dt, \quad f \in E, \quad \mu \in E^*. \tag{4.6.5}$$

Or, we may define, equivalently,

$$R_\lambda f = \int_0^\infty e^{-\lambda t}T_t f dt\left(=\lim_{n\to\infty}\frac{1}{n}\sum_{k=0}^{\infty}e^{-\lambda\frac{k}{n}}T_{\frac{k}{n}}f\right).$$

Since $T_t f$ is continuous and bounded as a function of t as we observed before, it is clear that these definitions make sense. Now

$$\|R_\lambda f\| \le \int_0^\infty e^{-\lambda t}\|T_t f\|dt \le \|f\|\int_0^\infty e^{-\lambda t}dt = \|f\|/\lambda.$$

Hence, we get

$$\|R_\lambda\| \le 1/\lambda \tag{4.6.6}$$

and thus R_λ is a bounded linear operator.

Next, let us show that

$$R_\lambda - R_\mu = -(\lambda - \mu)R_\lambda R_\mu \tag{4.6.7}$$

and

$$\|\lambda R_\lambda f - f\| \to 0, \quad \lambda \to \infty \tag{4.6.8}$$

hold. For $f \in E$, $\sigma \in E^*$, we have

$$\begin{aligned}(R_\lambda R_\mu f, \sigma) &= \int_0^\infty e^{-\lambda t}(T_t R_\mu f, \sigma)dt \\ &= \int_0^\infty e^{-\lambda t}(R_\mu f, T_t^* \sigma)dt \\ &= \int_0^\infty e^{-\lambda t} \int_0^\infty e^{-\mu s}(T_s f, T_t^* \sigma)dsdt \\ &= \int_0^\infty e^{-\lambda t} \int_0^\infty e^{-\mu s}(T_{s+t} f, \sigma)dsdt \\ &= \int_0^\infty e^{-\lambda t + \mu t} \int_t^\infty e^{-\mu s}(T_s f, \sigma)dsdt \\ &= \int_0^\infty e^{-\mu s}(T_s f, \sigma)ds \int_0^s e^{-(\lambda-\mu)t}dt \\ &= \int_0^\infty e^{-\mu s}(T_s f, \sigma)\frac{e^{-(\lambda-\mu)s} - 1}{-(\lambda-\mu)}ds \\ &= \frac{1}{-(\lambda-\mu)} \int_0^\infty (e^{-\lambda s} - e^{-\mu s})(T_s f, \sigma)ds \\ &= \frac{1}{-(\lambda-\mu)}((R_\lambda - R_\mu)f, \sigma),\end{aligned}$$

and

$$\|\lambda R_\lambda f - f\| \leq \int_0^\infty \lambda e^{-\lambda t}\|T_t f - f\|dt = \int_0^\infty e^{-s}\|T_{\frac{s}{\lambda}} f - f\|ds \to 0.$$

The range $\mathfrak{R}$ of R_λ is independent of λ due to (4.6.7). In fact, since

$$R_\mu f = R_\lambda[I + (\lambda-\mu)R_\mu]f \in \mathfrak{R}_\lambda,$$

we have $\mathfrak{R}_\mu \subset \mathfrak{R}_\lambda$. By interchanging λ and μ, we get $\mathfrak{R}_\lambda \subset \mathfrak{R}_\mu$. Therefore, we have $\mathfrak{R}_\lambda = \mathfrak{R}_\mu$, and we may simply write $\mathfrak{R}$ for $\mathfrak{R}_\lambda$.

$\mathfrak{R}$ is clearly a linear subspace, and by (4.6.8),

$$\overline{\mathfrak{R}} = E, \tag{4.6.9}$$

where $-$ indicates the norm closure in E. Thus, $\mathfrak{R}$ is a dense linear subspace of E.

Next, let us show that $\mathfrak{D}(A) = \mathfrak{R}$. First of all, let $f \in \mathfrak{R}$. Then, we can write $f = R_\lambda g$ for some $g \in E$, and hence

$$\begin{aligned}T_s f = T_s R_\lambda g &= \int_0^\infty e^{-\lambda t} T_{t+s} g dt \\ &= e^{\lambda s} \int_s^\infty e^{-\lambda t} T_t g dt \\ &= e^{\lambda s} \int_0^\infty e^{-\lambda t} T_t g dt - e^{\lambda s} \int_0^s e^{-\lambda t} T_t g dt, \\ \frac{T_s f - f}{s} &= \frac{e^{\lambda s} - 1}{s} R_\lambda g - e^{\lambda s}\frac{1}{s}\int_0^s e^{-\lambda t} T_t g dt \\ &\to \lambda f - g.\end{aligned}$$

Therefore, if $f \in \mathfrak{R}$, i.e., if $f = R_\lambda g$ for some g, then $f \in \mathfrak{D}(A)$ and

$$Af = \lambda f - g.$$

Next, we note that $\lambda - A$ is a one-to-one map. For this, it suffices to show that $(\lambda - A)f = 0 \Rightarrow f = 0$. From $(\lambda - A)f = 0$, we have $\lambda f = Af$; consequently,

$$T_s f = f + \lambda s f + o(s),$$
$$\|f\| \geq \|T_s f\| = (1 + \lambda s)\|f\| + o(s),$$
$$0 \geq \lambda s\|f\| + o(s), \quad \text{and therefore,} \quad \|f\| + o(1) \leq 0,$$

from which we obtain $\|f\| \leq 0$, and hence $\|f\| = 0$.

Let us now prove $f \in \mathfrak{D}(A) \Rightarrow f \in \mathfrak{R}$. If we can show this, we can conclude that $\mathfrak{D}(A) = \mathfrak{R}$. So, let $f \in \mathfrak{D}(A)$. Then, as Af exists, we can set $g = \lambda f - Af$ and let $f_0 = R_\lambda g$. From what we have already proved, it follows that

$$Af_0 = \lambda f_0 - g,$$

and therefore,

$$(\lambda - A)f_0 = g = (\lambda - A)f.$$

Since $\lambda - A$ is one-to-one, we get

$$f = f_0 = R_\lambda g \in \mathfrak{R}.$$

Summarizing what we obtained thus far, we see that $\mathfrak{D}(A) = \mathfrak{R}$, and if $f \in \mathfrak{D}(A)$ so that $f = R_\lambda g$, then

$$Af = \lambda f - g. \tag{4.6.10}$$

Furthermore, $\lambda - A$ is one-to-one, and

$$(\lambda - A)^{-1} = R_\lambda \text{ and hence } R_\lambda^{-1} = \lambda - A. \tag{4.6.11}$$

We also have $\overline{\mathfrak{D}(A)} = \overline{\mathfrak{R}} = E$.

If $f \in \mathfrak{D}(A)$, then $T_t f \in \mathfrak{D}(A)$, and the following evolution equation is valid:

$$\frac{dT_t f}{dt} = AT_t f \; (= T_t A f). \tag{4.6.12}$$

If $f \in \mathfrak{D}(A)$, then $f = R_\lambda g$ for some g, and

$$\begin{aligned} T_t f &= T_t R_\lambda g = R_\lambda T_t g \in \mathfrak{D}(A), \\ \frac{dT_t f}{dt} &= \lim_{\delta \downarrow 0} \frac{T_\delta - I}{\delta} T_t f = AT_t f = AT_t R_\lambda g \\ &= AR_\lambda T_t g = \lambda R_\lambda T_t g - T_t g \\ &= T_t(\lambda R_\lambda g - g) = T_t A R_\lambda g = T_t A f. \end{aligned}$$

4.7. Hille-Yosida Theory (ii). Construction of Semi-Group

In the preceding section, we investigated the properties of the generator of a semi-group and its relationship to the latter, assuming that a semi-group was given to start with. We have seen that if A is a generator, then

(A.1) A is a linear operator and $\overline{\mathfrak{D}(A)} = E$,

(A.2) $(\lambda - A)^{-1}$ is defined on all of E, and satisfies $\|(\lambda - A)^{-1}\| \leq 1/\lambda$ $(\lambda > 0)$.

It is the purpose of this section to prove that conversely a linear operator satisfying these conditions is the generator of a unique semi-group on E.

We shall first show that from the conditions above it follows that

$$\text{if we set } I_\lambda = \lambda(\lambda - A)^{-1}, \text{ then } \|I_\lambda f - f\| \to 0 \; (\lambda \to \infty) \tag{4.7.1}$$

holds.

Indeed, if $f \in \mathfrak{D}(A)$, then

$$
\begin{aligned}
I_\lambda f - f &= (\lambda - A)^{-1}\lambda f - (\lambda - A)^{-1}(\lambda - A)f \\
&= (\lambda - A)^{-1}(\lambda f - (\lambda - A)f) = (\lambda - A)^{-1}Af,
\end{aligned}
$$

from which it follows that

$$\|I_\lambda f - f\| \leq \|Af\|/\lambda \to 0 \ \ (\lambda \to \infty).$$

For an arbitrary $f \in E$, we can write, for any $\epsilon > 0$,

$$f = g + h, \quad g \in \mathfrak{D}(A), \quad \|h\| < \epsilon.$$

From the result above, we have $\|I_\lambda g - g\| \to 0$, and

$$
\begin{aligned}
\|I_\lambda f - f\| &\leq \|I_\lambda g - g\| + \|I_\lambda h\| + \|h\| \\
&\leq \|I_\lambda g - g\| + 2\|h\| < \|I_\lambda g - g\| + 2\epsilon \to 2\epsilon.
\end{aligned}
$$

Hence we have $\|I_\lambda f - f\| \to 0$ for an arbitrary $f \in E$.

As we remarked in the preceding section, $T_t = e^{tA}$ would be the semi-group whose generator is A, in case A is a bounded operator. In general, A is not bounded, so the story is not that simple. Instead of A, let us first consider $A_\lambda = AI_\lambda$. Since I_λ can be considered to be approximating operators for the identity operator I as we have seen above, A_λ may be regarded as approximating operators for A. Furthermore, since

$$
\begin{aligned}
A_\lambda = A \cdot \lambda \cdot (\lambda - A)^{-1} &= (\lambda - (\lambda - A))\lambda(\lambda - A)^{-1} \\
&= \lambda \cdot \lambda(\lambda - A)^{-1} - \lambda I = \lambda(I_\lambda - I),
\end{aligned}
$$

A_λ is bounded. If we now set

$$T_t^{(\lambda)} = e^{tA_\lambda}, \tag{4.7.2}$$

then these give semi-groups approximating T_t. Let us show that for each λ, $T_t^{(\lambda)}$ is a semi-group. Properties $T_t^{(\lambda)} \to I$, $T_t^{(\lambda)}T_s^{(\lambda)} = T_{t+s}^{(\lambda)}$ are clear. To see that $\|T_t^{(\lambda)}\| \leq 1$, we note

$$
\begin{aligned}
\|T_t^{(\lambda)}\| &= \|e^{tA_\lambda}\| = \|e^{t\lambda(I_\lambda - I)}\| = \|e^{-t\lambda}e^{t\lambda I_\lambda}\| \\
&\leq e^{-t\lambda}\|e^{t\lambda I_\lambda}\| \leq e^{-t\lambda}\sum_n \frac{\|t\lambda I_\lambda\|^n}{n!} \leq e^{-t\lambda}\sum \frac{(t\lambda)^n}{n!} = 1.
\end{aligned}
$$

We would like to obtain the desired semi-group T_t as

$$T_t = \lim_{\lambda \to \infty} T_t^{(\lambda)}, \tag{4.7.3}$$

but for this purpose we have to show that $T_t^{(\lambda)}$ actually converges as $\lambda \to \infty$.

First of all, note that

$$(\lambda - A)^{-1}(\mu - A)^{-1} = (\mu - A)^{-1}(\lambda - A)^{-1} \tag{4.7.4}$$

holds. In order to show this, it is enough to show that

$$(\lambda - A)(\mu - A)(\lambda - A)^{-1}(\mu - A)^{-1} = I,$$

but this can be done by a formal calculation, once we observe the fact that both of the ranges of $(\lambda - A)^{-1}$ and of $(\mu - A)^{-1}$ are contained in both of the domains of $\lambda - A$ and of $\mu - A$, because of the identities

$$\mu - A = (\mu - \lambda)I + (\lambda - A), \quad \lambda - A = (\lambda - \mu)I + (\mu - A).$$

From (4.7.4), it follows immediately that

$$I_\lambda I_\mu = I_\mu I_\lambda \text{ and hence } A_\lambda A_\mu = A_\mu A_\lambda \tag{4.7.5}$$

(recall $A_\lambda = \lambda(I_\lambda - I)$). Consequently, both $T_t^{(\lambda)}$ and $T_t^{(\mu)}$ commute with both A_λ and A_μ. Now, since $T_t^{(\lambda)} = e^{tA_\lambda}$, we have

$$\frac{dT_t^{(\lambda)} f}{dt} = A_\lambda T_t^{(\lambda)} f, \quad \frac{dT_t^{(\mu)} f}{dt} = A_\mu T_t^{(\mu)} f.$$

Hence, if we set $f_t = T_t^{(\lambda)} f - T_t^{(\mu)} f$, then

$$\frac{df_t}{dt} = A_\lambda T_t^{(\lambda)} f - A_\mu T_t^{(\mu)} f = A_\lambda f_t + g_t, \tag{4.7.6}$$

where

$$g_t = (A_\lambda - A_\mu) T_t^{(\mu)} f = T_t^{(\mu)} (A_\lambda f - A_\mu f). \tag{4.7.7}$$

If we solve the equation (4.7.6) in the same way as for a differential equation for a scalar-valued function, we obtain

$$\begin{aligned} f_t &= \int_0^t e^{(t-s)A_\lambda} g_s ds + f_0 = \int_0^t e^{(t-s)A_\lambda} g_s ds \\ &= \int_0^t T_{t-s}^{(\lambda)} g_s ds. \end{aligned}$$

Therefore, we have

$$\|f_t\| \le \int_0^t \|T_{t-s}^{(\lambda)} g_s\| ds \le \int_0^t \|g_s\| ds \le \|A_\lambda f - A_\mu f\| \cdot t.$$

Now if $f \in \mathfrak{D}(A)$, then

$$A_\lambda f - A_\mu f = (I_\lambda - I_\mu) A f \to 0 \quad (\lambda, \mu \to \infty),$$

and therefore, it follows from the inequality above that $f_t = T_t^{(\lambda)} f - T_t^{(\mu)} f$ tends to 0, as $\lambda, \mu \to \infty$, uniformly in t as long as t lies in a bounded interval (uniform convergence over compact sets). For an arbitrary f, we can approximate it arbitrarily closely by $g \in \mathfrak{D}(A)$, so that by noting that

$$\|T_t^{(\lambda)} f - T_t^{(\mu)} f\| \le 2\|f - g\| + \|T_t^{(\lambda)} g - T_t^{(\mu)} g\|,$$

we can conclude that $T_t^{(\lambda)} f$ converges, as $\lambda \to \infty$, uniformly in t over compact sets. Let us denote the limit by $T_t f$. It is easy to show that T_t is a semi-group on E. Let us show that the generator $\tilde{A}$ of T_t coincides with the original A.

For $f \in \mathfrak{D}(A)$, we have

$$T_t^{(\lambda)} f - f = \int_0^t T_s^{(\lambda)} A_\lambda f ds = \int_0^t T_s^{(\lambda)} I_\lambda A f ds,$$

and

$$\begin{aligned} \|T_s^{(\lambda)} I_\lambda A f - T_s A f\| &\le \|T_s^{(\lambda)} I_\lambda A f - T_s^{(\lambda)} A f\| + \|T_s A f - T_s^{(\lambda)} A f\| \\ &\le \|I_\lambda A f - A f\| + \|T_s A f - T_s^{(\lambda)} A f\| \\ &\to 0 \quad (\text{uniformly on } s \in [0, t]), \end{aligned}$$

from which it follows that

$$T_t f - f = \int_0^t T_s A f ds.$$

Therefore, we have

$$\frac{T_t f - f}{t} \to Af, \quad \text{i.e.,} \quad f \in \mathfrak{D}(\tilde{A}) \text{ and } \tilde{A}f = Af.$$

This implies that $\tilde{A} \supset A$. As we saw in the preceding section $(\lambda - \tilde{A})^{-1}$ exists, and by our hypothesis $(\lambda - A)^{-1}$ also exists. From $\tilde{A} \supset A$, it follows that $(\lambda - \tilde{A})^{-1} \supset (\lambda - A)^{-1}$, but since $(\lambda - A)^{-1}$ is an operator defined on all of E, we must have $(\lambda - \tilde{A})^{-1} = (\lambda - A)^{-1}$, from which it follows that $\tilde{A} = A$.

We have shown that there exists a semi-group T_t, which has A as its generator. In order to show the uniqueness of such a T_t, it suffices to show that if S_t is another semi-group with A as the generator, then $T_t f = S_t f$ holds. Since $\mathfrak{D}(A)$ is dense in E, and both T_t and S_t are bounded, it is enough to show that $T_t f = S_t f$ holds for $f \in \mathfrak{D}(A)$. For such an f, we have

$$\begin{aligned}\frac{d}{dt}(S_t f - T_t^{(\lambda)} f) &= AS_t f - A_\lambda T_t^{(\lambda)} f \\ &= A_\lambda (S_t f - T_t^{(\lambda)} f) + S_t (I - I_\lambda) A f.\end{aligned}$$

Solving this equation in the same way as we solved (4.7.6), we obtain

$$S_t f - T_t^{(\lambda)} f = \int_0^t T_{t-u}^{(\lambda)} S_u (I - I_\lambda) A f du.$$

Therefore, we have

$$\|S_t f - T_t^{(\lambda)} f\| \leq t \|(I - I_\lambda) A f\| \to 0, \quad \lambda \to \infty,$$

from which we conclude that

$$S_t f = T_t f$$

as desired.

REMARK. As it is evident from the proof above, it is not necessary to require that condition (A.2) holds for all $\lambda > 0$. If (A.2) is satisfied for some sequence $\lambda_n \to \infty$, then we can construct a semigroup T_t having A as the generator and it is unique.

4.8. Generators of Transition Probabilities (i). General Theory

Let $P(t, x, E)$ be the transition probability introduced in §4.4. As we explained in §4.5, to this transition probability there corresponds a semi-group T_t, $t > 0$, on the space $\boldsymbol{C}$. The generator A of this T_t, $t > 0$, will be called the generator of the transition probability $P(t, x, E)$. If $f \in \mathfrak{D}(A)$, then

(4.8.1)
$$Af(x) = \lim_{t \downarrow 0} \frac{\int_R f(y) P(t, x, dy) - f(x)}{t}.$$

We give some remarks concerning $\mathfrak{D}(A)$. The following three conditions are mutually equivalent.

(i) $f \in \mathfrak{D}(A)$.

(ii) $\varphi_t(x) \equiv \frac{1}{t}\left[\int_R f(y) P(t, x, dy) - f(x)\right]$ converges, as $t \to 0$, uniformly in x.

(iii) $\varphi_t(x)$ converges, as $t \to 0$, for each fixed x, and the limit is a continuous function of x.

The implications (i) $\Rightarrow$ (ii), (ii) $\Rightarrow$ (iii) are clear. Since $\varphi_t(x)$ is a continuous function of x for each t, the limit $\varphi(x)$ of $\varphi_t(x)$ is also continuous if condition (ii) is satisfied, and furthermore, $\|\varphi_t - \varphi\| \to 0$, and hence $f \in \mathfrak{D}(A)$. Finally, we derive (i) from (iii). For this purpose, let us denote by $\tilde{\mathfrak{D}}$ the set of all f which satisfies condition (iii), and for $f \in \tilde{\mathfrak{D}}$, define

$$\tilde{A}f(x) = \lim_{t \downarrow 0} \varphi_t(x).$$

Then, clearly, $\tilde{A}f \in \boldsymbol{C}$. If $f \in \mathfrak{D}(A)$, then $f \in \tilde{\mathfrak{D}}$ and $Af = \tilde{A}f$ holds, so that we have $A \subset \tilde{A}$. Next, we show that $\lambda - \tilde{A}$ is one-to-one. For this purpose, we first note that if $f(x)$ attains its minimum at $x = x_0$, then

$$\tilde{A}f(x_0) = \lim_{t \downarrow 0} \varphi_t(x_0) \geq 0$$

holds. In order to show that $\lambda - \tilde{A}$ is one-to-one, it suffices to show that

$$(\lambda - \tilde{A})u = 0 \Rightarrow u = 0.$$

Since u is a continuous function on a compact space, it attains the minimum value $u(x_0)$, and

$$u(x_0) = \frac{1}{\lambda}\tilde{A}u(x_0) \geq 0, \quad \text{and hence } u(x) \geq 0 \quad \text{always holds.}$$

Similarly, since $(\lambda - \tilde{A})(-u) = 0$, we obtain $-u(x) \geq 0$ also always holds, and therefore, we must have $u(x) \equiv 0$.

Now, since $A \subset \tilde{A}$, it follows that $\lambda - A \subset \lambda - \tilde{A}$. We know that $\lambda - A$ has the inverse $(\lambda - A)^{-1}$ $(= R_\lambda)$, while $(\lambda - \tilde{A})^{-1}$ also exists because of the observation made above, which clearly satisfies $R_\lambda = (\lambda - A)^{-1} \subset (\lambda - \tilde{A})^{-1}$. As R_λ is defined on all of $\boldsymbol{C}$, we conclude from the observations above that

$$(\lambda - \tilde{A})^{-1} = R_\lambda = (\lambda - A)^{-1} \quad \text{and hence} \quad \tilde{A} = A.$$

Therefore, $\tilde{\mathfrak{D}} = \mathfrak{D}(A)$. This completes the proof of (iii) $\Rightarrow$ (i).

In the preceding section, we showed that the conditions (A.1) and (A.2) are necessary and sufficient for an operator A on a Banach space to be a generator of a semi-group. Let us derive conditions necessary and sufficient for an operator A to be a generator of a transition probability $P(t, x, E)$, in particular.

THEOREM 4.8.1. *In order for an operator A to be a generator of a transition probability, it is necessary and sufficient that the following four conditions are satisfied:*

(a.1) *A is a linear operator on $\boldsymbol{C}$, and $\overline{\mathfrak{D}(A)} = \boldsymbol{C}$.*

(a.2) $A \cdot 1 = 0$.

(a.3) *If $u \in \mathfrak{D}(A)$ attains its minimum value at $x = x_0$, then $Au(x_0) \geq 0$.*

(a.4) *For any $\lambda > 0$, the equation $(\lambda - A)u = v$ always has a solution u for any $v \in \boldsymbol{C}$.*

PROOF. Since it is clear either from the definitions or from the general theory of semi-groups that these conditions are necessary, we will prove only the sufficiency. First of all, we show the fact that $(\lambda - A)^{-1}$ is defined on all of $\boldsymbol{C}$. Since we have condition (**a.4**), it suffices for this purpose to show that $\lambda - A$ is one-to-one, namely that $\lambda u = Au \Longrightarrow u = 0$. Since $u(x)$ is a continuous function, it attains its minimum value $u(x_0)$. Then,

$$u(x_0) = \frac{1}{\lambda} Au(x_0) \geq 0 \quad \text{(by } (\mathbf{a.3})\text{)},$$

from which it follows that $u(x) \geq 0$ always holds. As $\lambda(-u) = A(-u)$, we obtain similarly $-u(x) \geq 0$, and thus $u(x) \equiv 0$.

Next, we show that $\|(\lambda - A)^{-1}\| \leq 1/\lambda$. Let $u = (\lambda - A)^{-1}v$. If we let $u(x_0)$ be the minimum value of $u(x)$, then

$$v(x_0) = (\lambda - A)u(x_0) = \lambda u(x_0) - Au(x_0) \leq \lambda u(x_0),$$

from which it follows that

$$u(x) \geq u(x_0) \geq \frac{1}{\lambda} v(x_0) \geq -\frac{1}{\lambda}\|v\|.$$

Noting that $(-u) = (\lambda - A)^{-1}(-v)$, we also have

$$-u(x) \geq -\frac{1}{\lambda}\| - v\| = -\frac{1}{\lambda}\|v\|,$$

and we thus conclude that

$$|u(x)| \leq \frac{1}{\lambda}\|v\|,$$

and therefore,

$$\|u\| \leq \frac{1}{\lambda}\|v\|.$$

We now see that conditions (A.1) and (A.2) of the preceding section are satisfied, and hence A is the generator of some semi-group on $\boldsymbol{C}$.

Next we show that $T_t \geq 0$, i.e., if $u \geq 0$ (meaning that $u(x) \geq 0$ holds for every x), then $T_t u \geq 0$. To begin with we show that $(\lambda - A)^{-1} \geq 0$. For this it suffices to show that from $(\lambda - A)u = v, v \geq 0$, it follows that $u \geq 0$. So, if we let $u(x_0)$ be the minimum value of $u(x)$, then $Au(x_0) \geq 0$. Therefore, we have

$$\lambda u(x_0) = Au(x_0) + v(x_0) \geq 0,$$

from which it follows that

$$u \geq 0,$$

and therefore, $I_\lambda = \lambda(\lambda - A)^{-1} \geq 0$. We then obtain

$$T_t^{(\lambda)} = e^{tA_\lambda} = e^{t\lambda(I_\lambda - I)} = e^{-t\lambda}e^{tI_\lambda} = e^{-t\lambda}\sum \frac{(tI_\lambda)^n}{n!} \geq 0.$$

We thus obtain that

$$T_t = \lim_{\lambda\to\infty} T_t^{(\lambda)} \geq 0.$$

We now show that $T_t 1 = 1$. Since $(\lambda - A)\frac{1}{\lambda} = 1$ holds by $(\mathbf{a.2})$, we have

$$\frac{1}{\lambda} = (\lambda - A)^{-1}1 = \int_0^\infty e^{-\lambda t} T_t 1 dt,$$

from which it follows that $T_t 1 = 1$.

We note that for each fixed pair t, x, $T_t f(x)$ is a linear functional on f and satisfies

$$f \geq 0 \Rightarrow T_t f(x) \geq 0,$$
$$T_t 1(x) = 1.$$

Thus, by the Riesz representation theorem, there exists a probability distribution $P(t, x, E)$ such that

$$T_t f(x) = \int_R f(y) P(t, x, dy).$$

It is easy to show by using the fact that T_t, $t > 0$, is a semi-group that $P(t, x, E)$ satisfies the conditions for a transition probability. □

REMARK. It is enough to require that condition (**a.4**) be satisfied for some sequence $\lambda_n \to \infty$, as was the case for (A.2) in the preceding section.

4.9. Generators of Transition Probabilities (ii). Examples

EXAMPLE 4.9.1. We have already seen that when R is a finite set, then a semi-group P_t of matrices corresponds to a transition probability. $\boldsymbol{C}$ in this case is a vector space whose dimension equals the number of elements in the finite set R. If we designate the vectors by column vectors, then $T_t f$ coincides with $P_t f$, where the latter signifies the application of matrix P_t on the column vector f. Thus the semi-group P_t, $t > 0$, of matrices can be taken to be the semi-group associated with a transition probability. Consequently, the generator A can also be considered as a matrix and is given by

$$Af = \lim_{t \downarrow 0} \frac{P_t - I}{t} f, \quad f \in \mathfrak{D}(A).$$

Since $\boldsymbol{C}$ is a finite-dimensional vector space, from $\overline{\mathfrak{D}(A)} = \boldsymbol{C}$, we conclude that $\mathfrak{D}(A) = \boldsymbol{C}$ holds. Consequently, the equation above holds for an arbitrary $f \in \boldsymbol{C}$, and we have

$$A = \lim_{t \downarrow 0} \frac{P_t - I}{t}.$$

Since $P_t \geq 0$, which means that all the entries of P_t are ≥ 0, we have for $A = (a_{ij})$,

$$a_{ij} \geq 0 \quad (j \neq i). \tag{4.9.1}$$

Also, from $P_t \mathbf{1} = \mathbf{1}$, where $\mathbf{1}$ designates the vector whose components are all 1, we have $A\mathbf{1} = 0$, so that

$$\sum_j a_{ij} = 0. \tag{4.9.2}$$

These two conditions are clearly necessary for A to be the generator of the transition probability, but they are also sufficient. In order to show this, let us check that the conditions (**a.1**)–(**a.4**) of the preceding section are satisfied. It is clear that conditions (**a.1**) and (**a.2**) are satisfied. Now let u_{i_0} be the smallest component of a vector u. Then,

$$\begin{aligned}(Au)_{i_0} &= \sum_j a_{i_0 j} u_j = a_{i_0 i_0} u_{i_0} + \sum_{j \neq i_0} a_{i_0 j} u_j \\ &\geq a_{i_0 i_0} u_{i_0} + \sum_{j \neq i_0} a_{i_0 j} u_{i_0} \qquad \text{by (4.9.1)} \\ &= 0 \qquad \text{by (4.9.2)},\end{aligned}$$

which shows that (**a.3**) is satisfied. Finally, we show that for every v,

$$(\lambda - A) u = v \tag{4.9.3}$$

can be solved if λ is sufficiently large. We can get the solution u formally as

$$u = (\lambda - A)^{-1}v = \lambda^{-1}\left(1 - \frac{A}{\lambda}\right)^{-1} v = \lambda^{-1}\left(1 + \left(\frac{A}{\lambda}\right) + \left(\frac{A}{\lambda}\right)^2 + \cdots\right) v.$$

We see that the series on the right-hand side of the equation above converges if $\lambda > \|A\|$, and for such λ we can check that the u given by this convergent series solves equation (4.9.3), and therefore, condition (**a.4**) is satisfied.

EXAMPLE 4.9.2. Let us derive the generator for the transition probability given in Example 4.4.2. We shall use the following results obtained in §4.6:

$$\mathfrak{D}(A) = \mathfrak{R}, \quad AR_\lambda f = \lambda R_\lambda f - f.$$

In the sequel, the symbols $x, y, \cdots$ designate finite numbers. Since

$$T_t f(x) = \int N_t(y - x) f(y) dy, \quad T_t f(\infty) = f(\infty),$$

we have

$$R_\lambda f(x) = \int R_\lambda f(y - x) f(y) dy, \quad R_\lambda f(\infty) = \frac{1}{\lambda} f(\infty),$$

where

$$R_\lambda(x) = \int_0^\infty e^{-\lambda t} N_t(x) dt = \frac{1}{\sqrt{2\lambda}} e^{-\sqrt{2\lambda}|x|},$$

so that

$$R_\lambda f(x) = \int \frac{1}{\sqrt{2\lambda}} e^{-\sqrt{2\lambda}|y-x|} f(y) dy.$$

From the last equation, it follows that $R_\lambda f(x)$ is twice continuously differentiable and satisfies

$$(R_\lambda f)''(x) = 2\lambda R_\lambda f(x) - 2f(x),$$

and hence we have $\lim_{x\to\infty}(R_\lambda f)''(x) = 0$. From these observations we obtain

$$\mathfrak{D}(A) \subset \tilde{\mathfrak{D}} \equiv \{g/g \in \boldsymbol{C},\ g''(x) \text{ is continuous}, \lim_{x\to\infty} g''(x) = 0\},$$

and furthermore, for $g\,(= R_\lambda f) \in \mathfrak{D}(A)$, we have

$$Ag(x) = \lambda g(x) - f(x) = \frac{1}{2} g''(x), \quad Ag(\infty) = 0.$$

Now, let us consider the operator $\tilde{A}$, for which

$$\mathfrak{D}(\tilde{A}) = \tilde{\mathfrak{D}}, \quad \tilde{A}g(x) = \frac{1}{2} g''(x), \quad \tilde{A}g(\infty) = 0$$

are satisfied. Clearly, $\tilde{A} \supset A$ holds. Next, we show that $(\lambda - \tilde{A})^{-1}$ exists. For this, it suffices to show that from the assertion $\lambda u = \tilde{A}u$ follows that $u = 0$. Now the solution u of the equation $\lambda u = \tilde{A}u$ satisfies

$$\lambda u(x) = \frac{1}{2} u''(x),$$

so that we obtain

$$u(x) = ae^{\sqrt{2\lambda}x} + be^{-\sqrt{2\lambda}x}.$$

Since $u(x)$ is bounded, we must have $a = b = 0$, and hence $u = 0$, which implies that $(\lambda - \tilde{A})^{-1}$ exists. Consequently, we have

$$(\lambda - \tilde{A})^{-1} \supset (\lambda - A)^{-1} = R_\lambda, \text{ and therefore, } (\lambda - \tilde{A})^{-1} = (\lambda - A)^{-1},$$

and we conclude that $\tilde{A} = A$ holds.

EXAMPLE 4.9.3. For the case of Example 4.4.3, we get as for the preceding example that an element f of $\mathfrak{D}(A)$ is characterized by the conditions

$$\begin{aligned}
&f \in \boldsymbol{C}, \quad f''(x) \text{ is continuous } (0 < x < \infty), \\
&\lim_{x\to 0} f''(x) \text{ exists and is finite}, \quad \lim_{x\to\infty} f''(x) = 0, \\
&\lim_{x\to 0} f'(x) = 0
\end{aligned}$$

and satisfies

$$Af(x) = \frac{1}{2}f''(x), \quad Af(0) = \frac{1}{2}f''(0+), \quad Af(\infty) = 0.$$

EXAMPLE 4.9.4. Let R be the space of the real numbers mod 1. Then $\boldsymbol{C}$ becomes the space of all the periodic continuous functions on the real line with period 1. Let $\mathfrak{D}$ be the set of all the functions in $\boldsymbol{C}$ which are twice continuously differentiable, and for $f \in \mathfrak{D}$, let

$$Af = f'.$$

Let us show that this A is the generator of some transition probability. We do this by checking that A satisfies conditions (**a.1**)–(**a.4**) of the preceding section. It is clear that conditions (**a.1**)–(**a.3**) are satisfied. In order to show that A satisfies condition (**a.4**), it suffices to solve the equation

$$\lambda u - u' = v.$$

Here v is a periodic function of period 1, and we have to find a solution u also from functions having the same property. Multiplying both sides of the equation above by $e^{-\lambda x}$, we get

$$\begin{aligned}
\lambda e^{-\lambda x}u(x) - e^{-\lambda x}u'(x) &= e^{-\lambda x}v(x), \\
(-e^{-\lambda x}u(x))' &= e^{-\lambda x}v(x).
\end{aligned}$$

As $e^{-\lambda x}u(x) = 0$ at $x = \infty$, we get

$$\begin{aligned}
e^{-\lambda x}u(x) &= \int_x^\infty e^{-\lambda y}v(y)dy, \\
u(x) &= \int_x^\infty e^{-\lambda(y-x)}v(y)dy.
\end{aligned}$$

This function $u(x)$ indeed satisfies the equation $\lambda u - u' = v$, and since

$$\begin{aligned}
u(x+1) &= \int_{x+1}^\infty e^{-\lambda(y-x-1)}v(y)dy = \int_x^\infty e^{-\lambda(y-x)}v(y+1)dy \\
&= u(x) \quad (\text{note that } v(y+1) = v(y)),
\end{aligned}$$

$u(x)$ is a function of period 1.

Also we have

$$R_\lambda v(x) = \int_x^\infty e^{-\lambda(y-x)}v(y)dy = \int_0^\infty e^{-\lambda t}v(x+t)dt,$$

which implies that $T_t v(x) = v(x+t)$. Therefore, we have

$$P(t, x, E) = \delta(x+t, E),$$

where $x + t$ has to be considered, naturally, in the sense of mod 1.

EXAMPLE 4.9.5. Let R be the reals mod 1 as in the preceding example, and let $Au = u''/2$. This A also satisfies conditions (**a.1**)–(**a.4**), and hence, is the generator of some $P(t, x, E)$. Indeed, we can show that $\mathfrak{D}(A)$ is the totality of twice continuously differentiable functions, and $P(t, x, E)$ is given by the formula

$$P(t, x, E) = \int_E \tilde{N}_t(y - x)dy,$$

where

$$\tilde{N}_t(x) = \sum_{n=-\infty}^{\infty} N_t(x + 2n) = \frac{1}{\sqrt{2\pi t}} \sum_{n=-\infty}^{\infty} e^{-\frac{(x+2n)^2}{2t}}.$$

EXAMPLE 4.9.6. Let R be the same as in the preceding example, let $Au = u'''$, and let $\mathfrak{D}(A)$ be the totality of thrice continuously differentiable functions. Then, this A does not correspond to any transition probability on R. This is because condition (**a.3**) is not satisfied by A. In order to see this, it suffices to construct a thrice continuously differentiable periodic function u of period 1, which attains its minimum at 0, and satisfies $u'''(0) < 0$. It is clear that if such a function exists, then (**a.3**) fails to be valid. To obtain such a function, it is enough to take a function $u(x) \geq 0$, which is periodic with period 1, and has a power series expansion in the neighborhood of $x = 0$ in the form

$$u(x) = ax^2 + bx^3 + \cdots \quad \text{with } a > 0, b < 0.$$

For example,

$$u(x) = 2(\sin 2\pi x)^2 - (\sin 2\pi x)^3$$

has these properties and serves as a desired counter-example.

4.10. Markov Processes (i). Markov Property

Let $P(t, x, E)$ be a transition probability on R. The conditions to be satisfied by R and $P(t, x, E)$ are the same as those explained in §4.4. Intuitively speaking, the transition probability $P(t, x, E)$ gives the probability that an entity which was at x initially moves into the set E after the lapse of time t. In other words, it gives a probability law specifying the change of certain phenomena according to the lapse of time. Let us denote by $x(t)$ the state at the time t of this phenomena varying according to this probability law. It is clear that $x(t)$ should be a stochastic process, and hence, should be represented, from the view point of measure theoretical probability theory, as $x(t, \omega)$, $\omega \in \Omega(\boldsymbol{B}, P)$. If we now denote by Φ the probability distribution of the process at time $t = 0$, namely, the probability distribution of the initial state $x(0, \omega)$, then the probability law of $x(t, \omega)$ should be given naturally by

$$\begin{aligned} &P(x(0, \omega) \in E_0, x(t_1, \omega) \in E_1, x(t_2, \omega) \in E_2, \cdots, x(t_n, \omega) \in E_n) \\ &= \int_{E_0} \Phi(d\xi_1) \int_{E_1} P(t_1, \xi_1, d\xi_2) \int_{E_2} P(t_2 - t_1, \xi_2, d\xi_3) \\ &\qquad \cdots \int_{E_n} P(t_n - t_{n-1}, \xi_{n-1}, d\xi_n). \end{aligned} \tag{4.10.1}$$

In order to show that such a stochastic process $x(t, \omega)$ exists, it suffices to take $\Omega = R^{[0,\infty)}$ and $\omega = \prod_t \omega_t$, and let $x(t, \omega) = \omega_t$ and show that it is possible to introduce a probability P on Ω in such a way that (4.10.1) above is satisfied. The fact that this is possible can be shown in the same way as for the case where R is the real line R^1, by using the Kolmogorov Theorem. Since such an $x(t, \omega)$ is essentially unique, it is called the stochastic process determined by the transition probability $P(t,x,E)$ and the initial distribution Φ. This process is also called the **Markov process**, since it was first investigated by A. Markov. Although the probability law of $x(t,\omega)$ depends on the initial distribution Φ, it is sufficient to consider only the case where Φ is given by the δ-distribution $\delta(a,E)$, since the general case can be obtained by integrating the result for this special case by $\Phi(da)$. The case of the initial distribution being $\delta(a,E)$ means that $x(0,\omega)\equiv a$, which is the case where the process starts from the state a. Let us denote such a Markov process by $x^{(a)}(t,\omega)$. To be more precise, we should also denote $\omega^{(a)}$ instead of ω, but since this is clear from the context, we do not adopt this notation. From now on we call the system $x^{(a)}(t,\omega)$, $a\in R$, of Markov processes the Markov process corresponding to $P(t,x,E)$. The aim of probability theory is to investigate these $x^{(a)}(t,\omega)$'s, and the facts we collected up to now on transition probabilities and semi-groups are merely preparatory material for such investigations.

The most remarkable among the probabilistic properties enjoyed by the Markov process $x^{(a)}(t,\omega)$ is its Markov property, which we explain below.

(i) Let us denote by $\boldsymbol{B}_t$ (or by $\boldsymbol{B}_t^{(a)}$ if necessary) the smallest Borel field containing all the ω-sets of the form

$$\{\omega/x^{(a)}(s,\omega)\in E\},\quad s\le t,\quad E\in\boldsymbol{B}_R.$$

Then

$$\begin{aligned}P(x^{(a)}(t+u,\omega)\in E/\boldsymbol{B}_t)&=P(x^{(b)}(u,\omega)\in E)_{b=x^{(a)}(t,\omega)}\\&=P(u,x^{(a)}(t,\omega),E)\end{aligned}\tag{4.10.2}$$

holds with probability 1. This means, intuitively speaking, that once the state at the time t is fixed, then the process will go on as if it started from this state, independently of the development of the process up to the time t. We call this property the Markov property. In order to prove the relation stated above, it is sufficient to show that for $M\in\boldsymbol{B}_t$,

$$P(\{\omega/x^{(a)}(t+u,\omega)\in E\}\cap M)=E\{P(u,x^{(a)}(t,\omega),E);M\}\tag{4.10.3}$$

holds. When M is of the form

$$\{\omega/x^{(a)}(t_1,\omega)\in E_1,\cdots,x^{(a)}(t_n,\omega)\in E_n\},\qquad 0<t_1<t_2<\cdots<t_n<t,$$

(4.10.3) follows immediately from (4.10.1). Noting that both sides of equation (4.10.3) are additive in M, we can then prove that (4.10.3) holds for every $M\in\boldsymbol{B}_t$.

(ii) We can generalize the Markov property stated above to some extent in the following form: for $0<u_1<u_2<\cdots<u_m$ and $M\in\boldsymbol{B}_t$, it holds with probability 1 that

$$\begin{aligned}&P\{x^{(a)}(t+u_1)\in E_1,x^{(a)}(t+u_2)\in E_2,\cdots,x^{(a)}(t+u_m)\in E_m/\boldsymbol{B}_t\}\\&\quad=P\{x^{(b)}(u_1)\in E_1,x^{(b)}(u_2)\in E_2,\cdots,x^{(b)}(u_m)\in E_m\}_{b=x^{(a)}(t)}.\end{aligned}\tag{4.10.4}$$

This fact can be proved by using a similar argument as was used in proving (i).

(iii) Let us generalize (ii) above even further. For this purpose, we introduce a few notations. It is unnecessary, of course, to explain the meaning of $R^{[0,\infty)}$. $\xi\in R^{[0,\infty)}$ can be represented in the form $\prod_{0\le t<\infty}\xi_t$, and ξ_t is called the **t-th coordinate** of ξ and is written as $p_t(\xi)$. Let us denote by $\boldsymbol{B}(R^{[0,\infty)})$ the smallest Borel field containing all the subsets of $R^{[0,\infty)}$ of the form $p_t^{-1}(E)$, $E\in\boldsymbol{B}_R$. Now, by $x^{(a)}(\bullet,\omega)$, let us represent $\prod_u x^{(a)}(u,\omega)$, a random vector taking values in

$R^{[0,\infty)}$, whose u-th coordinate equals $x^{(a)}(u,\omega)$. Also, by $x^{(a)}(t+\bullet,\omega)$ we denote a similar random vector whose u-th coordinate is equal to $x^{(a)}(t+u,\omega)$. Then, for $\Xi \in \boldsymbol{B}(R^{[0,\infty)})$, it holds with probability 1 that

$$P\{x^{(a)}(t+\bullet) \in \Xi/\boldsymbol{B}_t\} = P\{x^{(b)}(\bullet) \in \Xi\}_{b=x^{(a)}(t)}. \tag{4.10.5}$$

If Ξ is of the form $p_{u_1}^{-1}(E_1) \cap p_{u_2}^{-1}(E_2) \cap \cdots \cap p_{u_m}^{-1}(E_m)$, then the equation above coincides with equation (4.10.4) of (ii). Observing that both sides of equation (4.10.5) are additive in Ξ, we can prove the assertion for the general case.

Corresponding to the properties described above, we have the following properties concerning conditional expectations.

(iv) For a bounded $\boldsymbol{B}_R$-measurable function f, we have

$$E\{f(x^{(a)}(t+u))/\boldsymbol{B}_t\} = E\{f(x^{(b)}(u))\}_{b=x^{(a)}(t)}. \tag{4.10.6}$$

If f is an indicator function of a $\boldsymbol{B}_R$-measurable set, then the assertion is clear from (i). For more general f, the assertion can be proved by noting that both sides of the equation above are linear in f.

(v) For a bounded $\boldsymbol{B}_R$-measurable function f of n-variables, we have

$$\begin{aligned} &E\{f(x^{(a)}(t+u_1), x^{(a)}(t+u_2), \cdots, x^{(a)}(t+u_n))/\boldsymbol{B}_t\} \\ &\quad = E\{f(x^{(b)}(u_1), x^{(b)}(u_2), \cdots, x^{(b)}(u_n))\}_{b=x^{(a)}(t)}. \end{aligned} \tag{4.10.7}$$

(vi) For a bounded measurable function f defined on $R^{[0,\infty)}(\boldsymbol{B}(R^{[0,\infty)}))$, we have

$$E\{f(x^{(a)}(t+\bullet))/\boldsymbol{B}_t\} = E\{f(x^{(b)}(\bullet))\}_{b=x^{(a)}(t)}. \tag{4.10.8}$$

(v) and (vi) can be derived from (ii) and (iii), respectively, in the same way as (iv) was derived from (i).

4.11. Markov Processes (ii). Properties of Sample Processes

We already discussed the notion of separability for the case of real-valued stochastic processes in §2.8. Since the stochastic processes we are dealing with here take values in a compact space R satisfying the second axiom of countability, we must give a new definition of the separability for our processes.

DEFINITION 4.11.1. We say that a stochastic process x_t, $t \in T$, taking values in R is **separable** if for an arbitrary real-valued continuous function f on R, the real-valued process $f(x_t)$, $t \in T$, is separable in the sense of §2.8.

We say that a system $\{f_\lambda\}$ of real-valued continuous functions on R is a **base** for the space of all the real-valued continuous functions on R, if an arbitrary real-valued continuous function can be approximated uniformly on R by finite linear combinations of elements from this set $\{f_\lambda\}$. From the fact that R is a compact space satisfying the second axiom of countability, it follows that there exists a countable base for real-valued continuous functions on R.

Although in the definition given above we required $f(x_t)$ to be separable for all real-valued continuous functions f on R, it is sufficient to require this to be the case for all f belonging to a base.

Using the fact that there exists a separable modification for a real-valued stochastic process, we can prove that a stochastic process taking values in R also has a separable modification. Therefore, we may assume that the Markov process defined in the preceding section is separable.

THEOREM 4.11.1. *For a sample process of a separable Markov process $x^{(a)}(t)$, both left and right limits exist at all t with probability* 1. *Furthermore, at each t, we have*

$$P(x(t+0) = x(t)) = 1. \tag{4.11.1}$$

PROOF. For $f \in \boldsymbol{C}$ satisfying $f \geq 0$, let

$$Y(t) = e^{-\lambda t} R_\lambda f(x^{(a)}(t)).$$

Then, we have

$$\begin{aligned}
Y(t) &= e^{-\lambda t} \int_0^\infty e^{-\lambda s} T_s f(x^{(a)}(t)) ds \\
&= e^{-\lambda t} \int_0^\infty e^{-\lambda s} E(f(x^{(b)}(s)))_{b=x^{(a)}(t)} ds \\
&= e^{-\lambda t} \int_0^\infty e^{-\lambda s} E\{f(x^{(a)}(t+s))/\boldsymbol{B}_t\} ds \\
&= \int_t^\infty e^{-\lambda s} E\{f(x^{(a)}(s))/\boldsymbol{B}_t\} ds,
\end{aligned}$$

and therefore,

$$\begin{aligned}
E\{Y(t+u)/\boldsymbol{B}_t\} &= \int_{t+u}^\infty e^{-\lambda s} E\{E\{f(x^{(a)}(s))/\boldsymbol{B}_{t+u}\}/\boldsymbol{B}_t\} ds \\
&= \int_{t+u}^\infty e^{-\lambda s} E\{f(x^{(a)}(s))/\boldsymbol{B}_t\} ds \quad (\text{as } \boldsymbol{B}_{t+u} \supset \boldsymbol{B}_t) \\
&\leq \int_t^\infty e^{-\lambda s} E\{f(x^{(a)}(s))/\boldsymbol{B}_t\} ds = Y(t).
\end{aligned}$$

This implies that $Y(t)$ is a super-martingale (i.e., $-Y(t)$ is a sub-martingale) with respect to $\boldsymbol{B}_t$, $t \geq 0$ (cf. §4.3). We also have

$$\begin{aligned}
E(Y(t)) &= e^{-\lambda t} E\{R_\lambda f(x^{(a)}(t))\} = e^{-\lambda t} \int_R R_\lambda f(x) P(t, a, dx) \\
&= e^{-\lambda t} T_t R_\lambda f(a),
\end{aligned}$$

which implies that $E(Y(t))$ is continuous in t. As $x^{(a)}(t)$ is separable, $R_\lambda f(x^{(a)}(t))$ is also separable, and hence so is $Y(t)$. Thus we can use Theorem 4.3.1 to conclude that

$$P(\mathfrak{A}(Y)) = 1,$$

where

$$\mathfrak{A}(Y) = \{\omega / Y(t-0, \omega) \text{ and } Y(t+0, \omega) \text{ exist at all } t\}.$$

Therefore, we have

$$P\{\mathfrak{A}(R_\lambda f(x^{(a)}(\bullet)))\} = 1.$$

Since $\lambda R_\lambda f(x)$ converges to $f(x)$ uniformly on R as $\lambda \to \infty$, we obtain further that

$$P\{\mathfrak{A}(f(x^{(a)}(\bullet)))\} = 1. \tag{4.11.2}$$

Suppose now that a sequence f_n, $n = 1, 2, \cdots$, of non-negative functions belonging to $\boldsymbol{C}$ separates points of R. Then the mapping

$$R \ni x \to \boldsymbol{f}(x) = (f_1(x), f_2(x), \cdots) \in \prod_n [0, \|f_n\|] \equiv K$$

is one-to-one and continuous with respect to the product topology on K. Since R is a compact Hausdorff space, the inverse mapping is also continuous, and therefore, this correspondence $x \leftrightarrow \boldsymbol{f}(x)$ is a homeomorphism. Now, since each f_n satisfies (4.11.2), and since the countable intersection of sets of probability 1 is also of probability 1, we have

$$P\left\{\bigcap_{n=1}^{\infty} \mathfrak{A}(f_n(x^{(a)}(\bullet)))\right\} = 1.$$

By the definition of the product topology on K, we then have

$$P\{\mathfrak{A}(\boldsymbol{f}(x^{(a)}(\bullet)))\} = 1.$$

Since $x \leftrightarrow \boldsymbol{f}(x)$ is a homeomorphism, we finally conclude that

$$P\{\mathfrak{A}(x^{(a)}(\bullet))\} = 1.$$

Turning to the proof of (4.11.1), observe that we can derive from (4.10.6) the identity

$$E\left\{f(x^{(a)}(t))g(x^{(a)}(t+u))\right\} = E\left\{f(x^{(a)}(t))T_u g(x^{(a)}(t))\right\},$$

holding for any pair $f, g \in \boldsymbol{C}$. By letting $u \downarrow 0$, we get

$$E\left\{f(x^{(a)}(t))g(x^{(a)}(t+0))\right\} = E\left\{f(x^{(a)}(t))g(x^{(a)}(t))\right\}.$$

We have, therefore, for any continuous function h on $R \times R$

$$E\left\{h(x^{(a)}(t), x^{(a)}(t+0))\right\} = E\left\{h(x^{(a)}(t), x^{(a)}(t))\right\},$$

which leads to the assertion (4.11.1) by taking h to be a bounded metric function of R. □

Suppose for a separable Markov process $x^{(a)}(t)$, we define

$$\tilde{x}^{(a)}(t) = x^{(a)}(t+0).$$

Then, due to the theorem above, a sample process of $\tilde{x}^{(a)}(t)$ is right continuous and has, at most, points of discontinuity of the first kind, and furthermore, $\tilde{x}^{(a)}(t)$ is equivalent to $x^{(a)}(t)$ in the weak sense (cf. §2.8). (t is called a **point of discontinuity of the first kind of f** if both $f(t-0)$ and $f(t+0)$ exist, and f is discontinuous at t.) Therefore, we may assume that a sample process of a Markov process is right continuous and has, at most, points of discontinuity of the first kind. We shall tacitly assume this fact in the sequel without stating it explicitly. If a sample process is continuous with probability 1, then a Markov process is called a **diffusion process**.

4.12. Markov Processes (iii). Strong Markov Property

We already explained the Markov property in §4.10. It was described as the property specifying that once the state of the process is known for some time moment t, then the probability law of the future change of the state of the process will be determined as if the process started from that state at the time t, independently of the development of the process up to the time t. In this description the time t was arbitrary, but had to be constant. Let us now investigate whether a similar property will hold if t is a non-constant random variable. This will not be the case for arbitrary random variables, but we shall show that it will hold for a special class of random variables $\tau(\omega)$, called Markov times.

DEFINITION 4.12.1. We say that $\tau = \tau^{(a)}(\omega)$ is a **Markov time** for a Markov process $x^{(a)}(t,\omega)$, $0 \leq t < \infty$, if τ is a random variable taking values in $[0, \infty]$, and satisfies the property that

$$\{\omega/\tau^{(a)}(\omega) < t\} \in \boldsymbol{B}_t \quad (\boldsymbol{B}_t^{(a)}, \text{ to be more precise}).$$

The possibility $\tau(\omega) = \infty$ is allowed. But if $P(\tau < \infty) = 1$ is satisfied, then it is called a **finite Markov time**.

We have defined $\boldsymbol{B}_t$ for an arbitrary constant t. We now extend this definition to the case of a Markov time τ as follows.

DEFINITION 4.12.2. Let τ be a Markov time for $x^{(a)}(t)$. Denote by $\boldsymbol{B}_\tau$ $(= \boldsymbol{B}_\tau^{(a)}$, to be more precise) the smallest Borel field containing all the ω-sets of the form $E_\alpha \cap \{\omega/\tau \geq \alpha\}$ $(\alpha > 0, E_\alpha \in \boldsymbol{B}_\alpha)$, and call it the Borel field determined by $x^{(a)}(t), t \leq \tau$.

(i) The simplest case of the strong Markov property can be stated as follows. If τ is a finite Markov time for $x^{(a)}(t)$, then we have

$$P(x^{(a)}(t+\tau) \in E/\boldsymbol{B}_\tau) = P(x^{(b)}(t) \in E)_{b=x^{(a)}(\tau)} = P(t, x^{(a)}(\tau), E)$$

and for a bounded $\boldsymbol{B}_R$-measurable function f

$$E\{f(x^{(a)}(t+\tau))/\boldsymbol{B}_\tau\} = E\{f(x^{(b)}(t))\}_{b=x^{(a)}(\tau)}.$$

It suffices to prove the latter identity for a continuous function f. In such a case, the right-hand side of the equation reduces to $T_t f(x^{(a)}(\tau))$. Thus, we need to show that for $M \in \boldsymbol{B}_\tau$,

$$E\{f(x^{(a)}(\tau+u)); M\} = E\{T_u f(x^{(a)}(\tau)); M\} \tag{4.12.1}$$

holds. Since both sides of the last identity are additive in M, it is enough to prove the identity for the case where $M = E_\alpha \cap \{\omega/\alpha \leq \tau < \beta\}$. Let us now put

$$\alpha_{n,i} = \alpha + \frac{i}{n}(\beta - \alpha), \quad \tau^{(n)} = \alpha_{n,i} \text{ (when } \alpha_{n,i-1} \leq \tau < \alpha_{n,i}).$$

Then, clearly, $\tau^{(n)} \downarrow \tau$, and hence by the right-continuity of $x^{(a)}(t)$ (cf. §4.11), we have

$$x^{(a)}(\tau+u) = \lim_{n\to\infty} x^{(a)}(\tau^{(n)}+u).$$

Now, let

$$\begin{aligned} A &= E\{f(x^{(a)}(\tau^{(n)}+u)); E_\alpha \cap \{\alpha \leq \tau < \beta\}\} \\ &= \sum_{i=1}^{n} E\{f(x^{(a)}(\alpha_{n,i}+u)); E_\alpha \cap \{\alpha_{n,i-1} \leq \tau < \alpha_{n,i}\}\}. \end{aligned}$$

Then we note that since $\alpha \leq \alpha_{n,i-1} < \alpha_{n,i}$, all of the sets E_α, $\{\omega/\tau < \alpha_{n,i-1}\}$, $\{\omega/\tau < \alpha_{n,i}\}$ belong to $\boldsymbol{B}_{\alpha_{n,i}}$, and hence the set $E_\alpha \cap \{\alpha_{n,i-1} \leq \tau < \alpha_{n,i}\}$ also belongs to $\boldsymbol{B}_{\alpha_{n,i}}$. Consequently, we have

$$A = \sum_{i=1}^{n} E\{E\{f(x^{(a)}(\alpha_{n,i}+u))/\boldsymbol{B}_{\alpha_{n,i}}\}; E_\alpha \cap \{\alpha_{n,i-1} \leq \tau < \alpha_{n,i}\}\},$$

which, in view of the Markov property,

$$
\begin{aligned}
&= \sum_{i=1}^{n} E\{T_u f(x^{(a)}(\alpha_{n,i})); E_\alpha \cap \{\alpha_{n,i-1} \le \tau < \alpha_{n,i}\}\} \\
&= E\{T_u f(x^{(a)}(\tau^{(n)})); E_\alpha \cap (\alpha \le \tau < \beta)\}.
\end{aligned}
$$

Since $T_t f(x)$ is continuous in x just as $f(x)$ is, we obtain the conclusion (4.12.1) by letting $n \to \infty$.

(ii) Generalizing the result above, we have the following result:

$$
\begin{aligned}
&P\{x^{(a)}(\tau + u_1) \in E_1, \cdots, x^{(a)}(\tau + u_n) \in E_n / \boldsymbol{B}_\tau\} \\
&= P\{x^{(b)}(u_1) \in E_1, \cdots, x^{(b)}(U_n) \in E_n\}_{b = x^{(a)}(\tau)}.
\end{aligned}
$$

Even more generally, we have for $\Xi \in \boldsymbol{B}(R^{[0,\infty)})$,

$$
P\{x^{(a)}(\tau + \bullet) \in \Xi / \boldsymbol{B}_\tau\} = P\{x^{(b)}(\bullet) \in \Xi\}_{b = x^{(a)}(\tau)},
$$

which can be stated in terms of conditional expectations as

$$
E\{f(x^{(a)}(\tau + \bullet)) / \boldsymbol{B}_\tau\} = E\{f(x^{(b)}(\bullet))\}_{b = x^{(a)}(\tau)},
$$

which holds for every bounded $\boldsymbol{B}(R^{[0,\infty)})$-measurable function f defined on $R^{[0,\infty)}$. In order to prove all these assertions, it suffices to prove the last one. If we note that both sides of the last equation are linear in f, we see that it is enough to prove the last equation for the case where f is a product of a finite number of continuous functions of a single variable. Let us consider here the case where f is a product of two such functions. A more general case can be proved by a similar argument.

So, what we have to show now is the following: for two continuous functions f_1 and f_2 on R and for $0 < u_1 < u_2$,

$$
E\{f_1(x^{(a)}(\tau + u_1)) f_2(x^{(a)}(\tau + u_2)) / \boldsymbol{B}_\tau\} = E\{f_1(x^{(b)}(u_1)) f_2(x^{(b)}(u_2))\}_{b = x^{(a)}(\tau)}.
$$

We can prove this by using essentially the same argument as was employed in proving assertion (i), but we should keep in mind that

$$
g_1(x) = T_{u_2 - u_1} f_2(x), \quad g_2(x) = f_1(x) g_1(x), \quad g_3(x) = T_{u_1} g_2(x)
$$

are all continuous functions, and that if $s_1 < s_2$, then we have $\boldsymbol{B}_{s_1} \subset \boldsymbol{B}_{s_2}$ so that

$$
E(E(\bullet / \boldsymbol{B}_{s_2}) / \boldsymbol{B}_{s_1}) = E(\bullet / \boldsymbol{B}_{s_1})
$$

holds true.

(iii) The following properties, which can be derived by using the strong Markov property, will be used quite often in the sequel. Let $\tau = \tau^{(a)}(\omega)$ be a finite Markov time for $x^{(a)}(t)$, and set

$$
\begin{aligned}
P^0(t, a, dy) &= P(x^{(a)}(t) \in dy, \ \tau > t), \\
T_t^0 f(a) &= \int_R f(y) P^0(t, a, dy), \\
R_\lambda^0 f(a) &= \int_0^\infty e^{-\lambda t} T_t^0 f(a) dt, \\
\varphi^{(a)}(ds, dy) &= P(\tau \in ds, x^{(a)}(\tau) \in dy), \\
\hat{\varphi}^{(a)}(\lambda, dy) &= \int_0^\infty e^{-\lambda s} \varphi(ds, dy).
\end{aligned}
$$

Then we have

$$R_\lambda f(a) = R_\lambda^0 f(a) + \int_{y\in R} R_\lambda f(y)\hat{\varphi}^{(a)}(\lambda, dy) \tag{4.12.2}$$

and

$$T_t f(a) = T_t^0 f(a) + \int_{y\in R}\int_{s=0}^{\infty} T_{t-s} f(y)\varphi^{(a)}(ds, dy). \tag{4.12.3}$$

For the proof, let $c_s(t)$ be the indicator function of the interval $[0, s)$. Then

$$\begin{aligned} E\{f(x^{(a)}(t)); \tau > t\} &= E\{f(x^{(a)}(t))c_\tau(t)\} \\ &= \int_R f(y)P^0(t, a, dy) = T_t^0 f(a), \end{aligned}$$

$$\begin{aligned} &E\left\{\int_0^\tau e^{-\lambda t} f(x^{(a)}(t))dt\right\} = E\left\{\int_0^\infty e^{-\lambda t} f(x^{(a)}(t))c_\tau(t)dt\right\} \\ &= \int_0^\infty e^{-\lambda t} E\{f(x^{(a)}(t))c_\tau(t)\}dt = \int_0^\infty e^{-\lambda t} T_t^0 f(a)dt = R_\lambda^0 f(a), \end{aligned}$$

and also

$$\begin{aligned} R_\lambda f(a) &= \int_0^\infty e^{-\lambda t} E\{f(x^{(a)}(t))\}dt \\ &= E\left\{\int_0^\infty e^{-\lambda t} f(x^{(a)}(t))dt\right\} \\ &= E\left\{\int_0^\tau e^{-\lambda t} f(x^{(a)}(t))dt\right\} + E\left\{\int_\tau^\infty e^{-\lambda t} f(x^{(a)}(t))dt\right\}. \end{aligned}$$

The first term of the right-hand side of the last identity equals $R_\lambda^0 f(a)$, while the second term equals

$$\begin{aligned} &E\left\{e^{-\lambda\tau}\int_0^\infty e^{-\lambda t} f(x^{(a)}(\tau + t))dt\right\} \\ &= E\left\{e^{-\lambda\tau} E\left\{\int_0^\infty e^{-\lambda t} f(x^{(a)}(\tau + t))dt/\boldsymbol{B}_\tau\right\}\right\} \\ &= E\left\{e^{-\lambda\tau}\int_0^\infty e^{-\lambda t} E\{f(x^{(a)}(\tau + t))/\boldsymbol{B}_\tau\}dt\right\} \\ &= E\left\{e^{-\lambda\tau}\int_0^\infty e^{-\lambda t} T_t f(x^{(a)}(\tau))dt\right\} \\ &= E\{e^{-\lambda\tau} R_\lambda f(x^{(a)}(\tau))\} \\ &= \int_{t=0}^\infty \int_{y\in R} e^{-\lambda t} R_\lambda f(y)\varphi^{(a)}(dt, dy) \\ &= \int_{y\in R} R_\lambda f(y)\hat{\varphi}^{(a)}(\lambda, dy), \end{aligned}$$

which proves assertion (4.12.2). If we take the Laplace transform with respect to t of the right-hand side of equation (4.12.3), then the result equals the right-hand side of (4.12.2), and hence equals $R_\lambda f(a)$. But since $R_\lambda f(a)$ is, by definition, the Laplace transform of $T_t f(a)$, we can conclude that the identity (4.12.3) must also hold.

4.13. Markov Times

Let us describe some of the properties of Markov time and give some important examples.

(i) If τ_1 and τ_2 are Markov times, then so are $\max(\tau_1, \tau_2)$ and $\min(\tau_1, \tau_2)$ because of the following relations:

$$\{\omega / \max(\tau_1, \tau_2) < t\} = \{\omega/\tau_1 < t\} \cap \{\omega/\tau_2 < t\},$$
$$\{\omega / \min(\tau_1, \tau_2) < t\} = \{\omega/\tau_1 < t\} \cup \{\omega/\tau_2 < t\}.$$

(ii) If $\tau_1 \leq \tau_2 \leq \cdots$ are all Markov times, then $\tau = \lim \tau_n$ is also a Markov time, because

$$\{\omega/\tau < t\} = \bigcup_n \bigcap_m \left\{\omega \Big/ \tau_m < t - \frac{1}{n}\right\}.$$

(iii) If $\tau_1 \geq \tau_2 \geq \cdots$ are all Markov times, so is $\tau = \lim \tau_n$, since

$$\{\omega/\tau < t\} = \bigcup_n \{\omega/\tau_n < t\}.$$

(iv) $\tau(\omega) \equiv t$ is a Markov time.

(v) **Dynkin's Lemma.** *If τ is a finite Markov time, then*

$$f(a) = -E \int_0^\tau Af(x^{(a)}(t))dt + Ef(x^{(a)}(\tau)), \quad f \in \mathfrak{D}(A).$$

PROOF. Let $h_\lambda = \lambda f - Af$. Then $f = R_\lambda h_\lambda$, and

$$\begin{aligned} f(a) &= R_\lambda h_\lambda(a) \\ &= \int_0^\infty e^{-\lambda t} T_t h_\lambda(a) dt \\ &= E\left\{\int_0^\infty e^{-\lambda t} h_\lambda(x^{(a)}(t))dt\right\} \\ &= E\left\{\int_0^\tau\right\} + E\left\{\int_\tau^\infty\right\} = A_\lambda + B_\lambda. \end{aligned}$$

Now, since $h_\lambda \to -Af$ as $\lambda \to 0$, we have

$$\begin{aligned} A_\lambda &\to -E \int_0^\tau Af(x^{(a)}(t))dt, \\ B_\lambda &= E\left\{e^{-\lambda\tau} \int_0^\infty e^{-\lambda t} h_\lambda(x^{(a)}(\tau + t))dt\right\} \\ &= E\left\{e^{-\lambda\tau} \int_0^\infty e^{-\lambda t} E(h_\lambda(x^{(a)}(\tau + t))/\boldsymbol{B}_\tau)dt\right\} \\ &= E\left\{e^{-\lambda\tau} \int_0^\infty e^{-\lambda t} T_t h_\lambda(x^{(a)}(\tau))dt\right\} \\ &= E\{e^{-\lambda\tau} R_\lambda h_\lambda(x^{(a)}(\tau))\} = E\{e^{-\lambda\tau} f(x^{(a)}(\tau))\} \\ &\to E\{f(x^{(a)}(\tau))\}. \end{aligned}$$

□

(vi) Let F be a closed subset of R. By the **first passage time** of F we mean the infimum of the time required for $x^{(a)}(t)$ to go outside of F, and we denote it by $\tau_F = \tau_F^{(a)}(\omega)$. At any time t before τ_F, $x^{(a)}(t)$ is in F, but for any ϵ, $x^{(a)}(t)$ goes outside of F at least once between τ_F and $\tau_F + \epsilon$. We do not know whether $x^{(a)}(t)$

is in F or not at the time τ_F. If $x^{(a)}(t)$ is continuous at $t = \tau_F$, then $x^{(a)}(\tau_F)$ lies on the boundary of F (which is contained in F since F is closed). τ_F is a Markov time since

$$\{\omega/\tau_F < t\} = \bigcup_{s<t}\{\omega/x^{(a)}(s,\omega) \in F^c\} = \bigcup_{r<t, r\in \boldsymbol{Q}} \{\omega/x^{(a)}(r,\omega) \in F^c\},$$

where $\boldsymbol{Q}$ denotes the set of all rational numbers. The second equality above holds since F^c is an open subset of R and $x^{(a)}(s,\omega)$ is right continuous in s. The set in the right-hand side of the equation above is a countable union of sets belonging to $\boldsymbol{B}_t$, and hence it also belongs to $\boldsymbol{B}_t$.

(vii) Let U be an open subset of R. We define the first passage time τ_U of U in the same way as we defined τ_F. Assume, in particular, that $x^{(a)}(t)$ is a diffusion process on R. Using the topological properties of R, we can find an increasing sequence of closed subsets of U such that

$$F_1 \subset F_2 \subset \cdots \subset F_n \subset \cdots \to U,$$

for which

$$\tau_{F_1} \le \tau_{F_2} \le \cdots \to \tau_U.$$

Hence by (vi) and (ii), we can conclude that τ_U is also a Markov time.

(viii) Let us consider $\tau = \tau_F^{(a)}(\omega)$ as a special case of (vi) in which F consists of the single point a. Then τ represents the time at which the sample process $x^{(a)}(t)$ which has started from a leaves a for the first time. Equivalently, we can say that τ represents the length of time during which $x^{(a)}(t)$ stays at a. Let us derive the distribution of τ. Let us set

$$p(t) = P(\tau > t).$$

Then, using the Markov property, we can obtain

$$p(t+s) = p(t)p(s).$$

In fact,

$$\begin{aligned} p(t+s) &= P(x^{(a)}(u) = a, 0 \le u \le t+s) \\ &= P(x^{(a)}(u) = a, 0 \le u \le t \text{ and } x^{(a)}(t+v) = a,\ 0 \le v \le s) \\ &= E\{P(x^{(a)}(t+v) = a, 0 \le v \le s/\boldsymbol{B}_t); x^{(a)}(u) = a, 0 \le u \le t\} \\ &= E\{P(x^{(b)}(v) = a, 0 \le v \le s)_{b=x^{(a)}(t)}; x^{(a)}(u) = a, 0 \le u \le t\} \\ &= E\{P(x^{(a)}(v) = a, 0 \le v \le s); x^{(a)}(u) = a, 0 \le u \le t\} \\ &= p(t)p(s). \end{aligned}$$

Therefore, it must be the case that either $p(t) \equiv 0$ or $p(t) > 0$ for all t. In the latter case, it must have the form $e^{-\lambda t}$ $(\lambda \ge 0)$, since $p(t) \le 1$. If $\lambda = 0$, then $p(t) \equiv 1$, while if $\lambda > 0$, then $p(t)$ decreases from 1 to 0 as t increases from 0 to ∞.

If $p(t) \equiv 0$, then the sample path $x^{(a)}(t)$ starting from a leaves a instantaneously. That is, no matter how small $\epsilon > 0$ is, there exists a time t, $0 < t < \epsilon$, at which $x^{(a)}(t)$ leaves a. In this case, we call a an **instantaneous state** or a **point of instantaneous holding**.

If $p(t) \equiv 1$, then the path $x^{(a)}(t)$ starting from a can never leave a. In this case, a is called a **trap**.

If $p(t) = e^{-\lambda t}$ $(\lambda > 0)$, then the distribution of τ is of the **exponential type**, and in this case τ is called **exponential holding time** or **exponential waiting time**, while a is called a **point of exponential holding**.

We next show that the following four conditions are mutually equivalent.

(A) Point a is a trap.

(B) $P(t, a, E) = \delta(a, E)$, $t > 0$.

(C) $T_t f(a) = f(a)$, $t > 0$, $f \in \boldsymbol{C}$.

(D) $Af(a) = 0$, $f \in \mathfrak{D}(A)$.

The implications (A) $\to$ (B) $\to$ (C) $\to$ (D) are evident. Let us suppose that (D) is satisfied. Then since $T_t f \in \mathfrak{D}(A)$ if $f \in \mathfrak{D}(A)$, we have

$$\frac{dT_t f(a)}{dt} = AT_t f(a) = 0,$$

from which it follows that $T_t f(a) = f(a)$ if $f \in \mathfrak{D}(A)$. Since $\overline{\mathfrak{D}(A)} = \boldsymbol{C}$, condition (C) follows immediately. The implication (C) $\to$ (B) is obvious. Finally, suppose (B) is satisfied. Then,

$$P(x^{(a)}(t) = a) = 1, \ \ t > 0.$$

Therefore, we have $P(x^{(a)}(t) = a, \ \text{for all rational } t > 0) = 1$. Since $x^{(a)}(t)$ is right continuous, we then have $P(x^{(a)}(t) = a, \ \text{for all real } t > 0) = 1$, i.e., a is a trap.

(ix) For a diffusion process, there exist no points of exponential holding.

PROOF. If a is not a trap, there exists $f \in \mathfrak{D}(A)$ such that $Af(a) \neq 0$. Furthermore, if τ is the time at which $x^{(a)}(t)$ leaves a for the first time, then τ is a finite Markov time. Since a sample process of a diffusion process is continuous, we have $x^{(a)}(\tau) = a$. Therefore, by Dynkin's Lemma in (v), we get

$$f(a) = -Af(a) \cdot E(\tau) + f(a),$$

and hence $E(\tau) = 0$. Therefore, $P(\tau = 0) = 1$ which implies that a is a point of instantaneous holding. □

(x) If a is an interior point of M (which is a closed or open set), then $E(\tau_M^{(a)}) > 0$.

PROOF. If $E(\tau_M^{(a)}) = 0$, then $P(\tau_M^{(a)} = 0) = 1$. Therefore,

$$P(x^{(a)}(0+) \notin M^i) = 1.$$

But since $P(x^{(a)}(0+) = a) = 1$, we have a contradiction to the hypothesis $a \in M^i$. □

(xi) If a is not a trap, then there exists a neighborhood U of a such that $E(\tau_U^{(b)})$ is uniformly bounded with respect to b. If the process is a diffusion process, in particular, then for any $\epsilon > 0$, we can find a suitable U for which $E(\tau_U^{(b)}) < \epsilon$ for all $b \in U$. However, this stronger result may not necessarily hold for Markov processes in general.

PROOF. Since a is not a trap, there exists $f \in \mathfrak{D}(A)$ such that $Af(a) \neq 0$. We may assume that $Af(a) > 0$, by replacing f by $-f$, if necessary. Since Af is continuous, we can find a positive number α and a neighborhood U of a such that $Af(b) > \alpha$, $b \in U$. By Dynkin's Lemma, we have

$$f(b) \leq -\alpha E(\tau_U^{(b)}) + \|f\|, \ \text{so that} \ E(\tau_U^{(b)}) \leq \frac{1}{\alpha}(\|f\| - f(b)) < \infty.$$

If the process is a diffusion process, in particular, we have $x^{(b)}(\tau_U^{(b)}) \in \overline{U}$. Since f is continuous, by choosing smaller U if necessary, we can assume that the variation of f within $\overline{U}$ is less than $\alpha\epsilon$. Therefore, we have $f(x^{(b)}(\tau_U^{(b)})) < f(b) + \alpha\epsilon$. Using Dynkin's Lemma again, we get

$$f(b) < -\alpha E(\tau_U^{(b)}) + (f(b) + \alpha\epsilon),$$

from which it follows that $E(\tau_U^{(b)}) < \epsilon$. The reason why this stronger version cannot hold in general can be seen in the case when a is a point of exponential holding. In such a case, no matter how small we take U to be, $\tau_U^{(a)}$ will be bigger than the holding time τ of a, and since $E(\tau) > 0$, we can never make $E(\tau_U^{(a)})$ less than the positive number $E(\tau)$. We will see later that there are Markov processes having points of exponential holding. □

4.14. Dynkin's Theorem on Generators

As we have already seen, the following quantities correspond to each other in a one-to-one fashion:

$$\text{transition probability } P(t,x,E) \longleftrightarrow \text{semi-group } T_t \longleftrightarrow \text{Markoff process } x^{(a)}(t).$$

The generator A of T_t is also called the generator of $P(t,x,E)$, as well as of $x^{(a)}(t)$. The following theorem due to Dynkin determines A in terms of $x^{(a)}(t)$.

THEOREM 4.14.1. *Let U be a neighborhood of a, let $\tau = \tau_U^{(a)}$, and define*

$$Df(U) = \frac{E(f(x^{(a)}(\tau))) - f(a)}{E(\tau)}.$$

(If $E(\tau) = \infty$, we define $Df(U) = 0$). Then, for every $f \in \mathfrak{D}(A)$, we have

$$Df(U) \to Af(a) \quad (U \to a). \tag{4.14.1}$$

Furthermore, if $Df(U)$ approaches a continuous function of a as $U \to a$, then $f \in \mathfrak{D}(A)$ and hence the relation (4.14.1) above holds. Here the statement $Df(U) \to Af(a)$ means more precisely that if we take U sufficiently small, then we have

$$|Df(U) - Af(a)| \le \sup_{b\in U} |Af(b) - Af(a)|. \tag{4.14.2}$$

PROOF. If a is a trap, then the assertion is clear, since the left-hand side of (4.14.2) equals 0 in this case. If a is not a trap, then τ is a finite Markov time so that by Dynkin's Lemma

$$f(a) = -E\int_0^\tau Af(x^{(a)}(t))dt + Ef(x^{(a)}(\tau)).$$

Since a is not a trap, we can make $E(\tau) < \infty$ by choosing U sufficiently small. Furthermore, since a is an interior point of U, we have $E(\tau) > 0$. Hence, from the last equation, we obtain

$$\begin{aligned}\left|\frac{Ef(x^{(a)}(\tau)) - f(a)}{E(\tau)} - Af(a)\right| &\le \frac{1}{E(\tau)} E\int_0^\tau |Af(x^{(a)}(t)) - Af(a)|dt \\ &\le \sup_{b\in U} |Af(b) - Af(a)| \to 0 \quad (U \to a).\end{aligned}$$

Let us now denote by $\tilde{\mathfrak{D}}$ the set of all f for which $Df(U)$ tends to a continuous function of a as $U \to a$, and define $\tilde{A}f$ to be $\lim_{U\to a} Df(U)$ for $f \in \tilde{\mathfrak{D}}$. Then it follows from what we obtained above that $\tilde{A} \supset A$ holds. Next, we show that $\lambda - \tilde{A}$ is one-to-one. For this purpose, it suffices to show that from the assertion $\tilde{A}u = \lambda u$ follows that $u = 0$. Suppose $u(x)$ takes its minimum value at a. Then from the definition of $\tilde{A}$ it follows that $\tilde{A}u(a) \geq 0$, and hence $\lambda u(a) \geq 0$. This implies that $u(a) \geq 0$, and therefore, $u(x) \geq 0$ for all x. Since $\tilde{A}(-u) = \lambda(-u)$, we can apply the same argument to conclude that $-u(x) \geq 0$ for all x as well. Thus, we must have $u(x) \equiv 0$ if $\tilde{A}u = \lambda u$, and we conclude that $(\lambda - \tilde{A})^{-1}$ exists. Since $\tilde{A} \supset A$, we have

$$(\lambda - \tilde{A})^{-1} \supset (\lambda - A)^{-1} = R_\lambda,$$

and since $\mathfrak{D}(R_\lambda) = \boldsymbol{C}$, we must have $(\lambda - \tilde{A})^{-1} = (\lambda - A)^{-1}$, and we obtain $\tilde{A} = A$, and $\tilde{\mathfrak{D}} = \mathfrak{D}(A)$. □

Next, let us try to rewrite $Df(U)$ above in a different form. Let us set

$$\pi_U(a, dy) = P(x^{(a)}(\tau) \in dy).$$

Then it is clear that the support of this measure $\pi_U(a, \bullet)$ is contained in U^c (it is contained in the boundary ∂U of U in case the process is a diffusion process).

$$Df(U) = \frac{\int_{U^c} \pi_U(a, dy) f(y) - f(a)}{E(\tau)}. \tag{4.14.3}$$

In order to rewrite $E(\tau)$ in a different form, let us introduce the following notation: $p_U(a) = E(\tau_U^{(a)})$. (We set $p_U(a) = 0$ if $a \in U^c$, since $\tau_U^{(a)} \equiv 0$ in this case.)

We shall first show that if $U \subset V$, then

$$p_V(a) = p_U(a) + \int \pi_U(a, dy) p_V(y) \tag{4.14.4}$$

holds.

For this purpose, observe that there exists a $\boldsymbol{B}(R^{[0,\infty)})$-measurable function ϕ_V defined on $R^{[0,\infty)}$ such that

$$\tau_V^{(a)} = \phi_V(x^{(a)}(\bullet)).$$

In order to see this, let us first define for a closed set F,

$$\phi_F(\xi) = \inf\{t / \xi(t) \in F^c, t \in \boldsymbol{Q} \cap [0, \infty)\}.$$

Then by the right continuity of $x^{(a)}(t)$ and the fact that F^c is an open set, we must have $\tau_F^{(a)} = \phi_F(x^{(a)}(\bullet))$. Now, for an open set V, we choose an increasing sequence of closed sets $\{F_n\}$ approximating V from inside, namely, $F_1 \subset F_2 \subset \cdots \to V$, and define

$$\phi_V(\xi) = \lim_{n \to \infty} \phi_{F_n}(\xi).$$

Then, we get $\tau_V^{(a)} = \phi_V(x^{(a)}(\bullet))$. Now, we can write

$$\begin{aligned}\tau_V^{(a)} &= \tau_U^{(a)} + \tau_V^{(a)} - \tau_U^{(a)} \\ &= \tau_U^{(a)} + \phi_V(x^{(a)}(\tau_U^{(a)} + \bullet)).\end{aligned}$$

Taking the strong Markov property into account, we then obtain

$$\begin{aligned} E(\tau_V^{(a)}) &= E(\tau_U^{(a)}) + E\{\phi_V(x^{(a)}(\tau_U^{(a)} + \bullet))\} \\ &= E(\tau_U^{(a)}) + E\{E\{\phi_V(x^{(a)}(\tau_U^{(a)} + \bullet))/\boldsymbol{B}_{\tau_U^{(a)}}\}\} \\ &= E(\tau_U^{(a)}) + E\{E\{\phi_V(x^{(b)}(\bullet))\}_{b=x^{(a)}(\tau_U^{(a)})}\} \\ &= E(\tau_U^{(a)}) + E\{E(\tau_V^{(b)})_{b=x^{(a)}(\tau_U^{(a)})}\}, \end{aligned}$$

from which we get

$$\begin{aligned} p_V(a) &= p_U(a) + E\{p_V(x^{(a)}(\tau_U^{(a)}))\} \\ &= p_U(a) + \int \pi_U(a, dy) p_V(y), \end{aligned}$$

which proves the identity (4.14.4). Since $E(\tau)$ appearing in (4.14.3) is nothing but $p_U(a)$, we have

$$E(\tau) = p_V(a) - \int \pi_U(a, dy) p_V(y),$$

so that we can now write

$$Df(U) = \frac{\int \pi_U(a, dy) f(y) - f(a)}{\int \pi_U(a, dy)(-p_V(y)) - (-p_V(a))}. \tag{4.14.5}$$

4.15. Examples of Markov Processes

Let us apply theoretical results obtained in the previous sections to the Markov processes which correspond to the transition probabilities introduced in §4.4 and §4.9.

EXAMPLE 4.15.1. **Markov process with finite states**. Let us consider the case of Example 4.4.1 (or Example 4.9.1). We use the same notation as before, and denote the points of R by $\{1, 2, \cdots, n\}$. A sample process of $x^{(i)}(t)$ must be a function with the initial value i and having only the first kind of discontinuity, and furthermore, taking values among $1, 2, \cdots, n$ only. Hence such a function must be a step function. Therefore, if a process starts from i, then it stays in i for a certain period of time τ, and then moves to some other state j, and after a certain period of stay in j, it moves again to another state k and so on. Since the set $\{i\}$ is an open set, we must have $E(\tau) > 0$. This means that i cannot be an instantaneous state. Thus, i must be either a point of exponential holding or a trap. If it is a trap, then since $Af(i) = 0$ regardless of what f we take, we must have

$$a_{ij} = 0, \quad j = 1, 2, \cdots, n.$$

If i is a point of exponential holding, then we have $P(\tau > t) = e^{-\lambda_i t}$ and $E(\tau) = 1/\lambda_i$ and we can show that λ_i and a_{ij} are related by

$$\lambda_i = \sum_{j \neq i} a_{ij} = -a_{ii}. \tag{4.15.1}$$

In order to show this, we note that it is clear from the definition of A that for sufficiently small t

$$p(t, i, i) = 1 + a_{ii} t + o(t) \tag{4.15.2}$$

holds. On the other hand, from the consideration of the meaning of τ, it follows that

$$p(t,i,i) = e^{-\lambda_i t} + \epsilon, \tag{4.15.3}$$

where ϵ denotes the probability that the process be back in i at time t after it has once left i. Since it is necessary that at least two jumps occur before time t in order for this to happen, the magnitude of ϵ is of order at most $O(t^2)$ (which can be verified rigorously by using the strong Markov property). Therefore, (4.15.3) is reduced to

$$p(t,i,i) = 1 - \lambda_i t + o(t). \tag{4.15.4}$$

Comparing this with (4.15.2), we obtain that $\lambda_i = -a_{ii}$. By the right continuity of $x^{(i)}(t)$, $x^{(i)}(\tau)$ must be in a new state other than the state i. Suppose this new state is $j \neq i$; then we can show that

$$P(x^{(i)}(\tau) = j) = \frac{a_{ij}}{\lambda_i}. \tag{4.15.5}$$

In order to see this, let us consider

$$\begin{aligned} &P\{x^{(i)}(\tau) = j, \tau < T\} \\ &= \sum_{k=1}^{n} P\left\{x^{(i)}(\tau) = j, \frac{k-1}{n}T \leq \tau < \frac{k}{n}T\right\}. \end{aligned}$$

Since $x^{(i)}(t)$ is a step function, the number of jumps which occur before time T is finite; hence, by taking n sufficiently large we can make the probability of having two jumps in any of the time intervals $[n^{-1}(k-1)T, n^{-1}kT]$, $k = 1, 2, \cdots, n$ as small as we desire. Therefore, the right side of the equality above equals

$$\sum_{k=1}^{n} P\left\{x^{(i)}(t) = i,\ t \leq \frac{k-1}{n}T \quad \text{and } x^{(i)}\left(\frac{k}{n}T\right) = j\right\} + o(1),$$

which can be rewritten by using the Markov property as

$$\sum_{k=1}^{n} P\left\{x^{(i)}(t) = i,\ t \leq \frac{k-1}{n}T\right\} P\left\{x^{(i)}\left(\frac{T}{n}\right) = j\right\} + o(1).$$

But we have

$$P\left(x^{(i)}\left(\frac{T}{n}\right) = j\right) = p\left(\frac{T}{n}, i, j\right) = a_{ij}\frac{T}{n} + o(n^{-1})$$

so that the last terms of the identity above are equal to

$$\begin{aligned} &\sum_{k=1}^{n} e^{-\lambda_i \frac{k-1}{n}T} a_{ij}\frac{T}{n} + n \cdot o(n^{-1}) + o(1) \\ &\to \int_0^T e^{-\lambda_i t} a_{ij}\,dt = \frac{a_{ij}}{\lambda_i}(1 - e^{-\lambda_i T}). \end{aligned}$$

By letting $T \to \infty$, we finally obtain (4.15.5).

Thus, the movement of $x^{(i)}(t)$ can be described as follows: it starts from i and after a period of exponential holding time (whose average equals $\lambda_i^{-1} = (\sum_{j\neq i} a_{ij})^{-1}$) it moves into j with the probability a_{ij}/λ_i. If $a_{ij} = 0$ $(j \neq i)$, then i is a trap. After it has moved into j, it goes through the same kind of behavior and then moves into another state k. It keeps on repeating the same kind of movement, but when it reaches a trap, then it stops there forever.

If we use Dynkin's Theorem, then the formulae (4.15.1) and (4.15.5) can be obtained immediately. As open sets U converging to i, we may take i itself ($\{i\}$, to be more precise). It is enough to consider the case where i is not a trap. Since we know that the holding time τ of i is exponential if i is not a trap, we let λ_i^{-1} be its average, i.e., we let the distribution of τ be $P(\tau > t) = e^{-\lambda_i t}$. If we denote by π_{ij} the probability $P(x^{(i)}(\tau) = j)$, then this corresponds to $\pi_U(a, dy)$ in Dynkin's Theorem. Therefore, we must have

$$Af(i) = \frac{\sum_{j \neq i} \pi_{ij} f(j) - f(i)}{\lambda_i^{-1}} = \sum_{j \neq i} \lambda_i \pi_{ij} f(j) - \lambda_i f(i).$$

On the other hand, from the definition of A, we have

$$Af(i) = \sum_{j \neq i} a_{ij} f(j) + a_{ii} f(i).$$

Comparing these two formulae, we obtain

$$\lambda_i = -a_{ii} \left(= \sum_{j \neq i} a_{ij} \right), \quad \pi_{ij} = \frac{a_{ij}}{\lambda_i},$$

which are nothing but (4.15.1) and (4.15.5), respectively.

EXAMPLE 4.15.2. **The Wiener Process**. Let $B(t, \omega)$ be a Wiener process and suppose $B(0, \omega) \equiv 0$. Define

$$x^{(a)}(t, \omega) = a + B(t, \omega), \quad x^{(\infty)}(t, \omega) \equiv \infty.$$

Then we obtain a family of stochastic processes defined on $R = R^1 \cup \{\infty\}$. This family is a Markov process (in the sense we defined in §4.10) having the transition probability of the form

$$P(t, x, E) = \int_E N_t(y - x) dy.$$

From now on we call this Markov process the Markov process derived from a Wiener process, or more simply, the Wiener process. This process is a diffusion process.

EXAMPLE 4.15.3 (A Wiener process having a reflecting barrier at 0). Take $x^{(a)}(t, \omega)$ of the preceding example and define

$$y^{(a)}(t) = |x^{(a)}(t)|, \quad a \in [0, \infty].$$

Then this gives us a Markov process corresponding to Example 4.4.3 (which is the same as Example 4.9.3). This is also a diffusion process.

EXAMPLE 4.15.4 (Rotation on the unit circle). Consider the motion of a particle which moves with the unit velocity in the positive (i.e., counter-clockwise) direction along the unit circle. We can describe points on the unit circle by real numbers mod 1, and the motion of the particle is described by

$$x^{(a)}(t, \omega) = a + t \pmod 1.$$

This gives us a Markov process corresponding to the transition probability

$$P(t, a, E) = \delta(a + t, E)$$

(cf. Example 4.9.4). This is also a diffusion process, but once the initial position a is given, $x^{(a)}(t,\omega)$ is determined independently of ω as $a+t$ (mod 1), and hence, this is a so-called **deterministic** process.

EXAMPLE 4.15.5. Take the Wiener process of Example 4.15.2 above, but consider it in the sense of mod 1. Then we get a Markov process on the unit circle. We call this process the Wiener process on the unit circle. This is also a diffusion process (cf. Example 4.9.5).

4.16. Temporally Homogeneous Additive Processes

In Example 4.15.2 of the preceding section we saw how a Wiener process can be considered as a Markov process on $R = R^1 \cup \{\infty\}$. This idea can be extended to more general temporally homogeneous additive processes.

Let $y(t,\omega)$, $0 \le t < \infty$, be a temporally homogeneous additive process. We suppose that $y(0,\omega) = 0$. Let Φ_t be the distribution of $y(t,\omega)$. Then for any u, it is also the distribution of $y(u+t,\omega) - y(u,\omega)$. From the properties of the additive process we have

$$\Phi_{t+s} = \Phi_t * \Phi_s. \tag{4.16.1}$$

Φ_t is infinitely divisible, and its characteristic function $\varphi_t(z)$ is given by

$$\begin{cases} \varphi_t(z) = e^{t\psi(z)}, \\ \psi(z) = imz - \dfrac{v}{2}z^2 + \displaystyle\int_{-\infty}^{\infty}\left(e^{izu} - 1 - \frac{izu}{1+u^2}\right)n(du), \end{cases} \tag{4.16.2}$$

where m is a real number, $v \ge 0$, and n is a measure on R^1 satisfying

$$\int_{-\infty}^{\infty} \frac{u^2}{1+u^2} n(du) < \infty. \tag{4.16.3}$$

Now, for $a \in R = R^1 \cup \{\infty\}$, let

$$x^{(a)}(t,\omega) = \begin{cases} a + y(t,\omega), & \text{if } a \in R^1, \\ \infty, & \text{if } a = \infty. \end{cases}$$

Then we get a Markov process with transition probability given by

$$P(t,a,E) = \Phi_t(E-a), \quad \text{where } E - a = \{\xi - a/\xi \in E\}. \tag{4.16.4}$$

The Chapman-Kolmogorov equation for the transition probability reduces to equation (4.16.1) above.

The semi-group corresponding to this Markov process is given by

$$T_t f(x) = \int \Phi_t(dy - x) f(y) = \int \Phi_t(dy) f(y+x). \tag{4.16.5}$$

The Fourier transform operator $\mathfrak{F}$ is defined by

$$\mathfrak{F}g(z) = \lim_{n\to\infty} \int_{-n}^{n} e^{izx} g(x)dx, \quad \mathfrak{F}d\mu(z) = \lim_{n\to\infty}\int_{-n}^{n} e^{izx} d\mu(x),$$

where the limits are taken in the sense of Schwartz' distribution. $\mathfrak{F}$ defined this way is the Fourier transform in the sense of Schwartz' distribution. If we apply $\mathfrak{F}$ to both sides of (4.16.5), then we get

$$\mathfrak{F}T_t f(z) = \varphi_t(-z)\mathfrak{F}f(z) = e^{t\psi(-z)}\mathfrak{F}f(z). \tag{4.16.6}$$

Thus, we see, by applying the Fourier transform, that this semi-group T_t has a very simple form. Let us now take the Laplace transform of both sides of the equation above. Since

$$\Re\psi(-z) = -\frac{v}{2}z^2 + \int_{-\infty}^{\infty} (\cos zu - 1)n(du) \leq 0,$$

the Lapalace transform on t of the right-hand side of the equation above exists. For the left-hand side, we assume that the Fourier transform $\mathfrak{F}$ and the Laplace transform on t commute. Then we get

$$\mathfrak{F}R_\lambda f(z) = \frac{1}{\lambda - \psi(-z)}\mathfrak{F}f(z). \tag{4.16.7}$$

Also, from (4.16.6) we get

$$\mathfrak{F}Af(z) = \psi(-z)\mathfrak{F}f(z),$$

i.e.,

$$\mathfrak{F}Af(z) = \left\{-imz - \frac{v}{2}z^2 + \int \left(e^{-izu} - 1 + \frac{izu}{1+u^2}\right) n(du)\right\} \mathfrak{F}f(z). \tag{4.16.8}$$

Properties of the Fourier transform give us

$$\mathfrak{F}f'(z) = -iz\mathfrak{F}f(z), \quad \mathfrak{F}f''(z) = -z^2\mathfrak{F}f(z),$$

and

$$\mathfrak{F}\Delta_u f(z) = (e^{-izu} - 1)\mathfrak{F}f(z), \quad \text{where} \quad \Delta_u f(x) = f(x+u) - f(x).$$

Using these, we formally apply the inverse Fourier transform to both sides of (4.16.8) to obtain

$$Af(x) = mf'(x) - \frac{v}{2}f''(x) + \int \left(f(x+u) - f(x) - \frac{u}{1+u^2}f'(x)\right) n(du). \tag{4.16.9}$$

The determination of $\mathfrak{D}(A)$, which is necessary for the precise definition of A, however, may be quite cumbersome.

4.17. Birth and Death Processes

Suppose in a colony of germs the number of germs in it changes according to the following law. The probability that a germ will be split into two within the duration of time dt is given by pdt, neglecting terms involving higher powers of dt, and the probability that it will die in the same duration of time is given by qdt. Furthermore, we assume that the death or split of different germs occur independently of one another. Now let us consider how many germs there would be at the time $t + dt$ if there are n germs at a certain time t. The probability of ℓ splits and m deaths having occurred during the time period dt is given by

$$\frac{n!}{\ell!m!(n-\ell-m)!}(pdt)^\ell(qdt)^m(1-(p+q)dt)^{n-\ell-m}.$$

It is sufficient to consider only the following cases:

(A) $\ell = 0, m = 0$,
(B) $\ell = 1, m = 0$,
(C) $\ell = 0, m = 1$,

since in all other cases the probability above will be given by a constant multiple of a power of dt higher than or equal to the square. The number of germs at the time

$t + dt$ will be n for case (A), $n+1$ for case (B), and $n-1$ for case (C). Therefore, the probability $P(dt, n, k)$ of changing from n to k in time dt is given by

$$\begin{cases} P(dt, n, n) = 1 - n(p+q)dt, \\ P(dt, n, n+1) = npdt, \\ P(dt, n, n-1) = nqdt, \end{cases} \tag{4.17.1}$$

where again terms involving higher powers of dt are neglected. Since the number of germs remains to be 0 once it reaches 0, we must also have

$$P(dt, 0, 0) = 1. \tag{4.17.2}$$

The stochastic process given this way is called a **birth and death process**. Actually, a more general process whose transition probability is given by generalizing (4.17.1) as follows is also called by the same name:

$$\begin{cases} P(dt, n, n) = 1 - (p_n + q_n)dt, \\ P(dt, n, n-1) = p_n dt, \\ P(dt, n, n-1) = q_n dt. \end{cases}$$

For instance, suppose it will become harder for germs to get nutrition as n increases, so that the probability of splitting gets smaller and that of death gets larger. Then we have a situation where

$$p_1 > \frac{p_2}{2} > \frac{p_3}{3} > \cdots, \quad q_1 < \frac{q_2}{2} < \frac{q_3}{3} < \cdots.$$

We call this process a **pure birth process** if $q_n \equiv 0$, and a **pure death process** if $p_n \equiv 0$.

Now if we try to discuss this stochastic process as a Markov process, which we have investigated thus far, we encounter one difficulty since the state space of this process is $\{0, 1, 2, \cdots\}$ and thus, is not compact. A naive approach would be to add a point ∞ and define

$$P(dt, \infty, \infty) = 1,$$

but this simple modification may turn out to be inappropriate. We will show later why this may be so, and also show how we can overcome this difficulty.

For the moment, we leave this difficulty as it is, and let us proceed a little further. As in the case of a Markov process with a finite number of states, which we have already discussed, we may describe a birth and death process in the following way. Suppose the process starts from the state n, and after an exponential holding time τ (whose average equals $(p_n + q_n)^{-1}$) it moves to either $n+1$ or to $n-1$ with the probability $p_n/(p_n + q_n)$ or $q_n/(p_n + q_n)$, respectively. After that it moves from the new state $n+1$ or $n-1$ in a similar manner.

Let us first consider the case of a pure death process. We assume in the sequel that q_n is strictly positive for all n. Then, the number of germs will continue to decrease until finally it reaches 0 and the germs will be extinguished. Let us determine the distribution of the **extinction time**. This is certainly a Markov time, but we will not be bothered with this fact now. If the process starts from n, then it moves $n \to n-1 \to \cdots \to 1 \to 0$; hence if we denote by τ_k the duration of stay in the state k, then the extinction time ϵ_n is given by

$$\epsilon_n = \tau_n + \tau_{n-1} + \cdots + \tau_1.$$

$\tau_n, \tau_{n-1}, \cdots, \tau_1$ are independent random variables, and have the exponential distribution $F_k(t) = 1 - e^{-q_k t}$ with the mean q_k^{-1}, $k = 1, 2, \cdots, n$, respectively. Hence the distribution of ϵ_n is given by the convolution of F_k's. The mean of ϵ_n is given by

$$E(\epsilon_n) = \sum_{k=1}^{n} E(\tau_k) = \sum_{k=1}^{n} q_k^{-1} < \infty,$$

and hence we have $P(\epsilon_n < \infty) = 1$, which implies that no matter how large the initial number of germs is, germs will be extinguished sooner or later.

In order to determine $P(t, n, k)$, let us denote by $F_{k,n}$ the convolution of the distributions $F_k, \cdots, F_n$. Clearly, $F_{k,n}$ gives the distribution of $\tau_n + \tau_{n-1} + \cdots + \tau_k$, and $F_{k,n}(t)$ represents the probability that the number of germs becomes less than k by the time t. Therefore, the probability that this number equals k at time t is given by

$$P(t, n, k) = F_{k+1,n}(t) - F_{k,n}(t), \quad k = 0, 1, \cdots, n-1, n,$$

where we set

$$F_{0,n}(t) = 0, \quad F_{n+1,n}(t) = 1.$$

Now, in order that $P(t, n, k)$ defined above can be extended to give a transition probability on the one point compactification R obtained by adjoining the point ∞ to the space $\{0, 1, 2, \cdots, n, n+1, \cdots\}$, we have to give appropriately the values of $P(t, \infty, k)$ for $k = 0, 1, 2, \cdots, \infty$. In defining these values, we have to keep in mind that for an arbitrary continuous function $f(n)$ on R, the function

$$T_t f(n) = \sum P(t, n, k) f(k)$$

should also be continuous. Since all the points of R except ∞ are isolated points, the continuity of a function defined on R means that the value of the function at ∞ must coincide with the limit as $n \to \infty$ of the value of the function at n. For instance, the function f_k defined for each $k \neq \infty$ by $f_k(m) = \delta_{mk}$ is continuous on R, and for this f_k, we have $T_t f_k(n) = P(t, n, k)$; hence we need to have

$$\lim_{n \to \infty} P(t, n, k) = P(t, \infty, k), \quad k \neq \infty.$$

In order to find $\lim_{n\to\infty} P(t, n, k)$, let us compute $\lim_{n\to\infty} F_{k,n}(t)$. As we remarked before, $F_{k,n}(t)$ is a distribution function of $\tau_n^{(n)} + \tau_{n-1}^{(n)} + \cdots + \tau_k^{(n)}$ (we added the superscript (n) to the τ's to indicate that they refer to the process starting from n), but since the distributions of $\tau_n^{(n+1)}, \tau_{n-1}^{(n+1)}, \cdots, \tau_k^{(n+1)}$ are the same as those of $\tau_n^{(n)}, \tau_{n-1}^{(n)}, \cdots, \tau_k^{(n)}$, respectively, the distribution of $\tau_n^{(n+1)} + \tau_{n-1}^{(n+1)} + \cdots + \tau_k^{(n+1)}$ is the same as the distribution of $\tau_n^{(n)} + \tau_{n-1}^{(n)} + \cdots + \tau_k^{(n)}$. Therefore, we have

$$\begin{aligned} F_{k,n+1}(t) &= P(\tau_{n+1}^{(n+1)} + \tau_n^{(n+1)} + \cdots + \tau_k^{(n+1)} \le t) \le P(\tau_n^{(n+1)} + \cdots + \tau_k^{(n+1)} \le t) \\ &= P(\tau_n^{(n)} + \cdots \tau_k^{(n)} \le t) = F_{k,n}(t). \end{aligned}$$

Therefore, $F_{k,n}(t)$ decreases as $n \to \infty$ so that it has the limit $G_k(t)$ (≥ 0), and we obtain

$$P(t, \infty, k) = G_{k+1}(t) - G_k(t),$$

from which we obtain

$$\begin{aligned}P(t,\infty,\infty) &= 1 - \sum_{k\neq\infty} P(t,\infty,k)\\ &= 1 - \lim_{m\to\infty}\sum_{k=0}^{m}(G_{k+1}(t)-G_k(t))\\ &= 1 - \lim_{m\to\infty} G_{m+1}(t).\end{aligned}$$

We now separate two cases.

(i) When $\sum_{k=1}^{\infty} q_k^{-1} = \infty$. Since $F_{k,n}(t) \downarrow G_k(t)$ (≥ 0), for $\lambda > 0$, we have

$$\begin{aligned}\int_0^\infty e^{-\lambda t} G_k(t)dt &= \lim_{n\to\infty}\int_0^\infty e^{-\lambda t} F_{k,n}(t)dt\\ &= \lim_{n\to\infty}\frac{1}{\lambda}\int_0^\infty e^{-\lambda t} dF_{k,n}(t).\end{aligned}$$

As the distribution $F_{k,n}(t)$ is the convolution of $F_k, F_{k+1}, \cdots, F_n$, the last term of the above equation equals

$$\begin{aligned}&\lim_{n\to\infty}\frac{1}{\lambda}\prod_{\nu=k}^{n}\int_0^\infty e^{-\lambda t} dF_\nu(t)\\ &= \lim_{n\to\infty}\frac{1}{\lambda}\prod_{\nu=k}^{n}\left(1+\frac{\lambda}{q_k}\right)^{-1} = \frac{1}{\lambda}\prod_{\nu=k}^{\infty}\left(1+\frac{\lambda}{q_k}\right)^{-1}.\end{aligned}$$

Because of the assumption that $\sum q_k^{-1} = \infty$, the infinite product appearing in the last term of the identity above equals 0, and we can conclude that $G_k(t) \equiv 0$, since $G_k(t)$ is monotone increasing in t just as each $F_{k,n}(t)$ is. We thus have

$$P(t,\infty,\infty) = 1, \quad \text{and hence} \quad P(t,\infty,k) = 0, \quad k \neq \infty.$$

This means that ∞ is a trap in this case, and the naive extension of $P(t,n,k)$ mentioned in the beginning of this discussion works in this situation.

(ii) When $\sum_{k=1}^{\infty} q_k^{-1} < \infty$. In this case the infinite product, which was 0 in the discussion of case (i) above, is always positive. Furthermore, it approaches 1 as $k \uparrow \infty$ so that we have

$$\lim_{k\to\infty}\int_0^\infty e^{-\lambda t} G_k(t)dt = \frac{1}{\lambda}.$$

Since $G_k(t)$ is monotone increasing in t as is each $F_{k,n}$, we have

$$\int_0^\infty e^{-\lambda t}\lim_{k\to\infty} G_k(t)dt = \frac{1}{\lambda},$$

from which we can conclude that

$$\lim_{k\to\infty} G_k(t) \equiv 1,$$

and hence that

$$P(t,\infty,\infty) = 0, \qquad \sum_{k\neq\infty} P(t,\infty,k) = 1.$$

Thus, the point ∞ is an instantaneous state, and hence it is important to describe how the process moves from ∞ to finite states. Therefore, in this case, the naive modification considered before is not applicable.

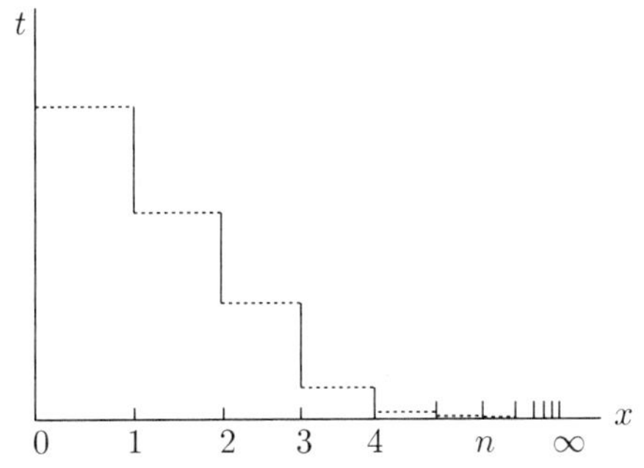

FIGURE 4.17

In both of these cases, there are no possibilities for the process to jump from finite states to the state ∞, but in case (i), the rate of decrease of the number of germs becomes more and more moderate as the initial number n of germs approaches ∞, and as a natural consequence, we have no other choice than to make the state ∞ to be a trap. On the other hand, in case (ii), the rate of decrease of the number of germs gets more and more drastic as n approaches ∞, so that even if the process reaches ∞ it must be brought back to finite states instantly. Since every finite state is always a point of exponential holding, a sample process starting from ∞ can be shown graphically as in Figure 4.17, where the points of R are taken on the abscissa and the time as the ordinate.

Next, we consider a pure birth process. We suppose that p_n is strictly positive for each $n \geq 1$. It is clear that 0 is a trap. If the process starts from n (≥ 1), then it keeps increasing thereafter. If we denote by τ_{nm} the time it takes for the process starting from n to reach m ($> n$), then similarly as before, we have

$$E(\tau_{nm}) = \sum_{\nu=n}^{m-1} p_\nu^{-1} < \infty.$$

Therefore, it is certain that the process reaches from n to m. If

$$\sum_{n=1}^{\infty} p_n^{-1} < \infty$$

is satisfied, then starting from 1 (hence starting from any n), the process goes, within a finite time τ, beyond any finite state, no matter how large this state is. Therefore, for any sufficiently large t, we must have

$$\sum_{m \neq \infty} P(t, n, m) < 1.$$

This means that if we want to define a Markov process on $R = \{1, 2, 3, \cdots, \infty\}$, then it is necessary to set

$$P(t, n, \infty) = 1 - \sum_{m \neq \infty} P(t, n, m).$$

Since $P(t, n, m) \equiv 0$ for $m < n$, we have to set $P(t, \infty, m) = \lim_{n \to \infty} P(t, n, m) = 0$, which means that ∞ becomes a trap. Furthermore, from a finite state n it reaches

∞ in a finite time τ_n, and

$$E(\tau_n) = \sum_{\nu=n}^{\infty} p_\nu^{-1} < \infty.$$

This τ_n is a Markov time, and is called the **explosion time**.

If, on the other hand,

$$\sum_{n=1}^{\infty} p_n^{-1} = \infty,$$

then

$$E\{e^{-\lambda \tau_{nm}}\} = \prod_{\nu=n}^{m-1} \int_0^\infty e^{-\lambda t} e^{-tp_\nu} p_\nu dt = \prod_{\nu=n}^{m-1} \left(1 + \frac{\lambda}{p_\nu}\right)^{-1}$$
$$\to 0 \quad (m \to \infty).$$

Therefore, if we define $\tau_n = \lim_{m\to\infty} \tau_{nm}$, then we have $E(e^{-\lambda\tau_n}) = 0$ so that $P(\tau_n = \infty) = 1$. This means that the process never explodes, and hence we have

$$\sum_{m \neq \infty} P(t, n, m) = 1.$$

The state ∞ is a trap for the same reason as before.

In the case of general birth and death processes, several interesting possibilities can occur depending on the relations between the p_n's and q_n's, but we omit their descriptions here.

CHAPTER 5

Diffusion

5.1. Diffusive Points

Let $x^{(a)}(t,\omega)$ be a Markov process on R. Notations and hypotheses used in this chapter will be the same as those in the preceding chapter. As we have already remarked, a sample process of $x^{(a)}(t,\omega)$ has the points of discontinuity of the first kind only. Let U be an open subset of R. We say that a Markov process $x^{(a)}(t,\omega)$ is **diffusive** in U if a process $x^{(a)}(t,\omega)$ starting from an arbitrary point $a \in U$ remains to be continuous up to the first passage time $\tau = \tau_U^{(a)}$ of U (including τ itself) with probability 1. By the assumption of continuity made above, $x^{(a)}(\tau_U^{(a)})$ lies on the boundary ∂U of U.

A point $b \in R$ is called a **diffusive point** if, in some neighborhood of b, the Markov process $x^{(a)}(t)$ is diffusive. The set of all diffusive points clearly forms an open set.

In general, a linear operator S is said to be **local** at a point x if, for a pair of functions $f, g \in \mathfrak{D}(S)$ satisfying $f = g$ in some neighborhood of x, the identity $Sf(x) = Sg(x)$ holds.

THEOREM 5.1.1. *The generator A of a Markov process $x^{(a)}(t,\omega)$ is local at every diffusive point of this process.*

PROOF. Let a be a diffusive point, and take a neighborhood U of a, in which $x^{(a)}(t)$ is diffusive. It is clear that in every neighborhood of a contained in U, $x^{(a)}(t)$ is diffusive. Take $f_1, f_2 \in \mathfrak{D}(A)$, and suppose that $f_1 = f_2$ holds in some neighborhood V of a. According to Dynkin's Theorem, we have

$$Af_i(a) = \lim_{W \downarrow a} \frac{Ef_i(x^{(a)}(\tau_W)) - f_i(a)}{E(\tau_W)}.$$

Whenever we take a neighborhood W of a in such a way that $\overline{W} \subset U \cap V$, then we have $x^{(a)}(\tau_W) \in U \cap V$ so that we have $Af_1(a) = Af_2(a)$. □

5.2. Ray's Theorem

Let $x^{(a)}(t)$ be a Markov process on R. We say that a point b of R has a **one-dimensional character** if there exists a neighborhood of b which is homeomorphic to an interval on the real line. This property has nothing to do with the process $x^{(a)}(t)$, and depends only on R. The set of all points having a one-dimensional character is clearly an open set. We say that a point b of R is a **one-dimensional diffusive point** for $x^{(a)}(t)$ if b is a point of a one-dimensional character as well as a diffusive point for $x^{(a)}(t)$. The set of all one-dimensional diffusive points is also an open set.

Let b be a diffusive point, and let U be a neighborhood of b. From the fact that b is a diffusive point, we may expect that the probability $P(t,b,U^c)$ for a Markov process $x^{(b)}(t)$ starting from b to go outside of U is very small if t is sufficiently small. The theorem of Ray which we will explain in this section will tell us that this probability $P(t,b,U^c)$ is indeed of the order of $o(t)$.

THEOREM 5.2.1. *If b is a one-dimensional diffusive point for $x^{(a)}(t)$, then for any neighborhood U of b,*

$$P(t,b,U^c) = o(t) \tag{5.2.1}$$

holds.

PROOF. By the assumption, $x^{(a)}(t)$ is diffusive in some neighborhood of b. When a neighborhood U of b gets smaller, $P(t,b,U^c)$ gets larger, and therefore, we may assume that $x^{(a)}(t)$ is diffusive in U. As b has a neighborhood homeomorphic to a line segment, we may represent points of some neighborhood of b as points on a line segment, namely by using real numbers. Likewise, we may represent U as an interval (u_1, u_2).

Now suppose the assertion (5.2.1) is not valid. Then, there exist $t_n \downarrow 0$ and $c > 0$ such that

$$P(t_n, b, U^c) > ct_n \tag{5.2.2}$$

for each n. Denote by $\tau_1^{(b)}$ the time when the process $x^{(b)}(t)$ leaves U for the first time through the end point u_1. We set $\tau_1^{(b)} = \infty$ if either $x^{(b)}(t,\omega)$ stays within U forever, or it goes out of U for the first time through the other end point u_2. To give a more rigorous definition, let $\tau_U^{(b)}$ be the first passage time for U. If $\tau_U^{(b)} < \infty$, then from the fact that $x^{(a)}(t)$ is diffusive in U, it follows that $x^{(b)}(\tau_U^{(b)})$ must be a boundary point of U, i.e., u_1 or u_2. So, we define

$$\tau_1^{(b)} = \begin{cases} \tau_U^{(b)}, & \text{when } x^{(b)}(\tau_U^{(b)}) = u_1, \\ \infty, & \text{otherwise.} \end{cases}$$

We define $\tau_2^{(b)}$ likewise by using u_2 in place of u_1. From (5.2.2) it follows that

$$P(\tau_1^{(b)} \le t_n) + P(\tau_2^{(b)} \le t_n) \ge P(x^{(b)}(t_n) \in U^c) > ct_n.$$

Therefore, either

$$P(\tau_1^{(b)} \le t_n) > \frac{c}{2} t_n$$

happens for infinitely many n, or

$$P(\tau_2^{(b)} \le t_n) > \frac{c}{2} t_n$$

happens for infinitely many n. We want to deduce a contradiction from each of these cases, but since the argument is exactly the same, we will show that the latter case leads to a contradiction. Let us use c for $c/2$ above and write the condition as

$$P(\tau_2^{(b)} \le t_n) > ct_n, \quad c > 0, \quad t_n \downarrow 0, \tag{5.2.3}$$

from which we will obtain a contradiction.

If we take a point y between b and u_2, we have

$$P(\tau_2^{(b)} \le t_n) \le P(\tau_2^{(b)} - \tau_2^{(b)}(y) \le t_n, \tau_2^{(b)}(y) \le t_n),$$

where $\tau_2^{(b)}(y)$ is defined in the same way as $\tau_2^{(b)}$ by using the interval (u_1, y) instead of $U = (u_1, u_2)$. By using the strong Markov property, we can rewrite the right-hand side of the inequality above as

$$\begin{aligned}
&E\{P(\tau_2^{(b)} - \tau_2^{(b)}(y) \le t_n / \boldsymbol{B}_{\tau_2^{(b)}(y)}); \tau_2^{(b)}(y) \le t_n\} \\
&= E\{P(\tau_2^{(y)} \le t_n); \tau_2^{(b)}(y) \le t_n\} \\
&\le P(\tau_2^{(y)} \le t_n),
\end{aligned}$$

where $\tau_2^{(y)}$ is defined in the same way as $\tau_2^{(b)}$ for $U = (u_1, u_2)$ by using y instead of b. Therefore, we obtain, by using (5.2.3), that for each $b \le y < u_2$,

$$P(\tau_2^{(y)} \le t_n) > ct_n, \quad c > 0, \quad t_n \downarrow 0. \tag{5.2.4}$$

Now, let $\epsilon = (u_2 - b)/4\ (> 0)$, $a = b + 2\epsilon$ and define

$$y(t, \omega) = \begin{cases} x^{(a)}(t, \omega), & t < \tau_U^{(a)}, \\ x^{(a)}(\tau_U^{(a)}, \omega), & t \ge \tau_U^{(a)}. \end{cases}$$

Then, a sample process of $y(t)$ is continuous, and always stays within the interval $[u_1, u_2]$. Therefore, if we take s sufficiently small, then we have

$$P(a - \epsilon < y(t) < a + \epsilon,\ 0 \le t < s) > \frac{1}{2}. \tag{5.2.5}$$

Since a sample process of $y(t)$ is uniformly continuous for $0 \le t \le s$, we have

$$\begin{aligned}
\alpha_n &= P(y((k-1)t_n) < a + \epsilon,\ y(kt_n) = u_2 \text{ hold for some } k \text{ with } kt_n \le s) \\
&\to 0 \quad (n \to \infty).
\end{aligned} \tag{5.2.6}$$

On the other hand, we also have

$$\alpha_n \ge \sum_{k=1}^{[s/t_n]} P(a - \epsilon < y(t) < a + \epsilon, 0 \le t \le (k-1)t_n \text{ and } y(kt_n) = u_2).$$

Since $y(t)$ is $\boldsymbol{B}_t$-measurable, the right-hand side of the inequality above equals

$$\begin{aligned}
&\sum_{k=1}^{[s/t_n]} E\{P(y(t_n) = u_2 / \boldsymbol{B}_{(k-1)t_n}); a - \epsilon < y(t) < a + \epsilon, 0 \le t \le (k-1)t_n\} \\
&= \sum E\{P(\tau_2^{(b)} \le t_n)_{b = x^{(a)}((k-1)t_n)}; a - \epsilon < y(t) < a + \epsilon, 0 \le t \le (k-1)t_n\}.
\end{aligned}$$

Now, by the definition of $y(t, \omega)$, $x^{(a)}((k-1)t_n)$ coincides with $y((k-1)t_n)$ on the set $\{a - \epsilon < y(t) < a + \epsilon, 0 \le t \le (k-1)t_n\}$, and therefore, $x^{(a)}((k-1)t_n)$ lies inside of $(a - \epsilon, a + \epsilon)$. As $(a - \epsilon, a + \epsilon) \subset [b, u_2)$, we see that from (5.2.4) it follows that $P(\tau_2^{(b)} \le t_n)_{b = x^{(a)}((k-1)t_n)} > ct_n$. Therefore, we can conclude that

$$\begin{aligned}
\alpha_n &\ge \sum_k ct_n P(a - \epsilon < y(t) < a + \epsilon, 0 \le t \le (k-1)t_n) \\
&\ge \frac{1}{2} ct_n [s/t_n] \ge \frac{c}{4} s,
\end{aligned}$$

which contradicts (5.2.6). □

THEOREM 5.2.2. *Suppose an open subset U of R is homeomorphic to an interval, and suppose further that every point of U is a diffusive point. If F is a closed subset of U, then*

$$P(t, b, U^c) = o(t) \tag{5.2.7}$$

holds uniformly in $b \in F$.

PROOF. If the assertion were true with smaller U and larger F, then the assertion of the theorem would certainly hold. Therefore, we may assume that every point of $\overline{U}$ is a one-dimensional diffusive point, and that F is homeomorphic with a closed interval J. Let us write $J = [b_1, b_2]$. Applying the preceding theorem, we have

$$P(t, b_i, U^c) = o(t), \quad i = 1, 2, \tag{5.2.8}$$

where we may assume that the term $o(t)$ can be chosen independently of i. By using the strong Markov property, we have for $b_1 \le b \le b_2$,

$$\begin{aligned} P(t, b, U^c) &= \int_0^t \varphi_1(ds) P(t-s, b_1, U^c) + \int_0^t \varphi_2(ds) P(t-s, b_2, U^c) \\ &= o(t), \end{aligned}$$

where φ_i denotes the distribution of the time when the process $x^{(b)}(t)$ leaves the interval (b_1, b_2) through the end point b_i. □

If we reexamine the proof of Theorem 5.2.1, we see that we have actually proved the following fact:

$$P(\tau_1^{(b)} < t) + P(\tau_2^{(b)} < t) = o(t).$$

If we denote by $Q(t, b, U^c)$ the probability that the process $x^{(b)}(s)$ is in the set U^c for some $0 \le s \le t$, then we see that the left-hand side of the equation above is equal to $Q(t, b, U^c)$. Hence, we also have the following:

THEOREM 5.2.3. *The assertions of the preceding two theorems are valid for $Q(t, b, U^c)$ as well.*

5.3. Local Generators

We say that a Markov process $x^{(a)}(t)$ has a local character at a point b, if for an arbitrary neighborhood U of b

$$P\{x^{(b)}(t) \in U^c\} = P(t, b, U^c) = o(t) \tag{5.3.1}$$

holds. Ray's Theorem in the preceding section tells us that $x^{(a)}(t)$ has a local character at each of its 1-dimensional diffusive points.

We assume in the sequel that the process has a local character at b. We say that a function $f \in \mathfrak{D}(A_b)$ if f is continuous and bounded in some sufficiently small neighborhood V of b, and if

$$\lim_{t \downarrow 0} \frac{1}{t} \left\{ \int_V P(t, b, dy) f(y) - f(b) \right\} \tag{5.3.2}$$

exists. We denote this limit by $A_b f$. From the fact that $x^{(a)}(t)$ has a local character at b, it follows that this definition is independent of the choice of V. It is clear that A_b thus defined has the following properties.

$(A_b.1)$ (Local character). If $f \in \mathfrak{D}(A_b)$ and if $f = g$ holds in some neighborhood of b, then $g \in \mathfrak{D}(A_b)$ and

$$A_b f = A_b g$$

holds.

$(A_b.2)$ (Linearity). If $f, g \in \mathfrak{D}(A_b)$, then $\alpha f + \beta g \in \mathfrak{D}(A_b)$ and

$$A_b(\alpha f + \beta g) = \alpha A_b f + \beta A_b g$$

holds.

$(A_b.3)$ (Positivity). If $f \geq f(b)$ holds in some neighborhood of b and if $f \in \mathfrak{D}(A_b)$, then $A_b f \geq 0$ holds.

$(A_b.4)$ If $f \in \mathfrak{D}(A)$, then $f \in \mathfrak{D}(A_b)$ and $A_b f = Af(b)$ holds.

With these properties of A_b in mind, we now define the local generator A_U in the following way.

DEFINITION 5.3.1. Suppose the process has a local character at every point of an open set U. Let

$$\mathfrak{D}(A_U) = \{f/f \in \mathfrak{D}(A_b) \text{ for every } b \in U, \text{ and } A_b f \text{ is continuous in } b \in U\},$$
$$A_U f(b) = A_b f, \quad b \in U.$$

We call this operator A_U the **local generator** of $x^{(a)}(t)$ in U.

As we pointed out in the beginning of this section, a process has a local character at its one-dimensional diffusive points. Therefore, we can consider the quantity A_b at such points. In fact, A_b at such points can also be defined in the following way.

DEFINITION 5.3.2. Define

$$P_U(t, a, E) = P\{\omega/x^{(a)}(t) \in E \text{ and } x^{(a)}(s) \in U, 0 \leq s \leq t\},$$

$$A_b f = \lim_{t \downarrow 0} \frac{1}{t} \left\{ \int_U P_U(t, b, dy) f(y) - f(b) \right\}.$$

In fact, this second definition of A_b turns out to be more useful than the first one.

The fact that both definitions coincide can be seen in the following way:

$$\begin{aligned} &\left| \int_U P(t, b, dy) f(y) - \int_U P_U(t, b, dy) f(y) \right| \\ &\leq \int_U (P(t, b, dy) - P_U(t, b, dy)) |f(y)| \\ &\leq \sup_U |f| \cdot P\{x^{(b)}(s) \notin U \text{ for some } 0 \leq s \leq t\}. \end{aligned}$$

But the probability appearing in the last term of the inequality above is exactly $Q(t, b, U^c)$ defined in the preceding section and Theorem 5.2.3 tells us that it is of the order $o(t)$.

5.4. Classification of One-Dimensional Diffusive Points

The set of all one-dimensional diffusive points is an open set and can be represented as a union of at most countable connected components. Furthermore, each component is homeomorphic to an open interval on the real line. Take one such component I and suppose I is homeomorphic to an open interval (r_1, r_2). Then points of I can be represented by real numbers in (r_1, r_2).

Let $b \in I$. Then the probability that $x^{(b)}(t)$ is continuous for $0 \le t < \tau_I^{(b)}$ is 1. Let us call a point b a **right shunt** if

$$P(x^{(b)}(t) \ge b \text{ for every } t \text{ such that } 0 \le t < \tau_I^{(b)}) = 1. \tag{5.4.1}$$

Let U be a neighborhood of b contained in I. Since $\tau_U^{(b)} \le \tau_I^{(b)}$ always holds, we must have, if b is a right shunt,

$$P(x^{(b)}(t) \ge b \text{ for every } t \text{ such that } 0 \le t < \tau_U^{(b)}) = 1. \tag{5.4.2}$$

Conversely, we can show that the validity of this condition for some neighborhood $U \subset I$ implies that b is a right shunt. Let us prove this fact. Let us denote by τ the first time the process $x^{(b)}(t)$ goes out of $[b, r_2]$ through b. We set $\tau = \infty$ if this does not occur. τ thus defined is a Markov time. If we set $\tau_n = \min(\tau, n)$, then for each n, τ_n is a finite Markov time. What we have to prove is $P(\tau < \infty) = 0$. Since

$$P(\tau < \infty) = \lim_{n\to\infty} P(\tau < n) = \lim_{n\to\infty} P(\tau_n < n),$$

it is enough to show that $P(\tau_n < n) = 0$. From the definition of τ it follows that

$$\begin{aligned}
0 &= P(\tau < n \text{ and } x^{(b)}(\tau_n + t) \ge b \text{ for sufficiently small } t) \\
&= P(\tau_n < n \text{ and } x^{(b)}(\tau_n + t) \ge b \text{ for sufficiently small } t) \\
&= E\{P\{x^{(b)}(t) \ge b \text{ for sufficiently small } t/\boldsymbol{B}_{\tau_n}\}; \tau_n < n\} \\
&= E\{P\{x^{(b)}(t) \ge b \text{ for sufficiently small } t\}; \tau_n < n\} \\
&\ge E\{P\{x^{(b)}(t) \ge b \text{ for } 0 \le t < \tau_U^{(b)}\}; \tau_n < n\} \\
&= P\{\tau_n < n\} \quad \text{(by the hypothesis (5.4.2))}.
\end{aligned}$$

Replacing the inequality $x^{(b)}(t) \ge b$ by $x^{(b)}(t) \le b$ in (5.4.1), we define a **left shunt**. The same assertion as (5.4.2) is valid for left shunts, and also the corresponding condition for any neighborhood U $(\subset I)$ of b would imply that b is a left shunt.

A right shunt, which is at the same time a left shunt, is actually a trap. A right shunt which is not a trap is called a **proper right shunt**. A **proper left shunt** is defined similarly. A point in I which is neither a right nor a left shunt is called a **regular point**. We will denote by $\Lambda_\ell, \Lambda_r, \Lambda_{p\ell}, \Lambda_{pr}, \Lambda_2$, and Λ_t the set of all left shunts, right shunts, proper left shunts, proper right shunts, regular points, and traps, respectively.

When we say that I is homeomorphic to some open interval, we should note that there actually exist several homeomorphisms between I and this open interval. We classify these homeomorphisms into two groups according to their orientation properties. Namely, we fix an arbitrary pair of points in I, and classify homeomorphisms depending on which of these two points would be mapped under the homeomorphism concerned onto the larger of the corresponding two real numbers. This classification does not depend on the choice of the pair of points in I. For two homeomorphisms of the same class, the definition of left and right shunts would be the same, while for homeomorphisms of different classes, the definition of left and right would be reversed. However, the definitions of traps and regular points remain the same no matter which homeomorphism we choose.

(i) If b is a right shunt, then the probability that $x^{(a)}(t)$ crosses b from right to left is 0.

PROOF. Let τ be the time when the process $x^{(a)}(t)$ crosses b from right to left for the first time. Then τ is a Markov time (set $\tau = \infty$ if this does not occur). Using an argument similar to the one used for deriving (5.4.1) from (5.4.2), we can prove that $P(\tau < \infty) = 0$. □

(ii) Suppose for every a satisfying $b < a < r_2$ the condition

$$P\{x^{(a)}(t) \geq b \text{ holds for every } t \text{ in } 0 \leq t < \tau_I^{(a)}\} = 1 \tag{5.4.3}$$

is satisfied. Then b is a right shunt.

PROOF. From the assumption it follows that if we fix t, then for $b < a < r_2$ we have

$$P\{\tau_I^{(a)} \leq t \quad \text{or} \quad b \leq x^{(a)}(t) < r_2 \quad \text{hold}\} = 1. \tag{5.4.4}$$

Now, choose u_2 in (b, r_2), v_1 in (r_1, b), and v_2 in (b, u_2). Then, by Ray's Theorem, there exists $\delta(t)$ which converges to 0 as $t \to 0$ for which

$$P\{x^{(a)}(s) \in (r_1, u_2) \quad \text{for} \quad 0 \leq s < t\} > 1 - \delta(t) \cdot t$$

is satisfied. $\delta(t)$ above can be chosen to be the same for all a in $v_1 \leq a \leq v_2$. In particular, for a satisfying $b < a \leq v_2$, we have from the assumption (5.4.3) that

$$P\{x^{(a)}(s) \in [b, u_2) \quad \text{for} \quad 0 \leq s < t\} > 1 - \delta(t) \cdot t.$$

Therefore, we also have

$$P\{x^{(a)}(t) \in [b, u_2)\} > 1 - \delta(t) \cdot t, \quad \text{i.e.,}$$
$$P(t, a, [b, u_2)) > 1 - \delta(t) \cdot t.$$

Since $P(t, a, \bullet) \to P(t, b, \bullet)$ (vague convergence), as $a \downarrow b$, we obtain

$$P(t, b, [b, r_2)) \geq 1 - \delta(t) \cdot t, \quad \text{i.e.,}$$
$$P\{x^{(b)}(t) \in [b, r_2)\} \geq 1 - \delta(t) \cdot t,$$

which clearly implies that

$$P\{\tau_I^{(b)} \leq t \quad \text{or} \quad b \leq x^{(b)}(t) < r_2 \quad \text{hold}\} \geq 1 - \delta(t) \cdot t. \tag{5.4.5}$$

From (5.4.4), we also have, for every a in $b < a < r_2$,

$$P\{\tau_I^{(a)} \leq t \quad \text{or} \quad b \leq x^{(a)}(t) < r_2\} \geq 1 - \delta(t) \cdot t. \tag{5.4.6}$$

Now, if we can show for any fixed t,

$$P\{x^{(b)}(s) \in [b, r_2) \quad \text{for every } s \text{ in } 0 \leq s < \min(\tau_I^{(b)}, t)\} = 1 \tag{5.4.7}$$

holds, then we can conclude, by letting $t \uparrow \infty$, that b is a right shunt. To obtain (5.4.7), it is enough to show, keeping in mind that $x^{(b)}(s)$ is continuous in $0 \le s < \tau_I^{(b)}$, that the following holds:

$$P\left\{x^{(b)}\left(\frac{k}{n}t\right) \in [b, r_2) \text{ for all } k\ (0 \le k \le n) \text{ such that } \frac{k}{n}t < \tau_I^{(b)}\right\}$$
$$\to 1 \quad (n \to \infty).$$

From the Markov property, (5.4.5), and (5.4.6), it follows that

$$\text{the probability in question above} \ge \left(1 - \delta\left(\frac{t}{n}\right) \cdot \frac{t}{n}\right)^n > 1 - \delta\left(\frac{t}{n}\right) \cdot t \to 1. \quad \square$$

(iii) Λ_r is a closed subset of R.

PROOF. We suppose that $b_n \in \Lambda_r$, $b_n \to b$, and show that $b \in \Lambda_r$. Without loss of generality, we may suppose that $b_n \uparrow b$ or $b_n \downarrow b$.

Suppose $b_n \uparrow b$. Using the result of (i), we get from the fact $b_n \in \Lambda_r$ that

$$P\{x^{(b)}(t) \in [b_n, r_2) \quad \text{for } t \text{ in } 0 \le t < \tau_I^{(b)}\} = 1.$$

Letting $n \to \infty$, we get

$$P\{x^{(b)}(t) \in [b, r_2) \quad \text{for } t \text{ in } 0 \le t < \tau_I^{(b)}\} = 1,$$

which means that

$$b \in \Lambda_r.$$

Suppose $b_n \downarrow b$. Again by (i), the fact that $b_n \in \Lambda_r$ implies that for any $a \in (b_n, r_2)$,

$$P\{x^{(a)}(t) \in [b_n, r_2) \quad \text{for } t \text{ in } 0 \le t < \tau_I^{(a)}\} = 1,$$

and therefore, that

$$P\{x^{(a)}(t) \in [b, r_2) \quad \text{for } t \text{ in } 0 \le t < \tau_I^{(a)}\} = 1.$$

Since $b_n \downarrow b$, this must hold for any $a > b$. Therefore, by (ii) we can conclude that $b \in \Lambda_r$. $\square$

(iv) The set Λ_ℓ is also closed. Therefore, Λ_t is closed as well, and Λ_2 is an open set.

5.5. Feller's Canonical Scale

As we remarked in the preceding section, the set of all one-dimensional diffusive points is an open set, each component of which is homeomorphic to an open interval on the real line. Therefore, we shall represent points of a component by points of the corresponding open interval (r_1, r_2). Now if we consider only regular points in the component, then the set of such points is again an open set in (r_1, r_2), and hence can be represented as a union of at most countable disjoint open sub-intervals of (r_1, r_2). Let us take one such interval or open sub-interval of such an interval and again call it $I = (r_1, r_2)$. By definition, all the points of $I = (r_1, r_2)$ are regular. We call such an interval a **regular interval**.

Take an open sub-interval $J = (j_1, j_2)$ of I in such a way that $\overline{J} \subset I$. Then, all the points including the end points of J are regular points. Denote by $s(a; j_2, j_1)$ the probability for a process $x^{(a)}(t)$ starting from a point a in $\overline{J}$ to reach j_2 before it reaches j_1, and by $s(a; j_1, j_2)$ the probability of reaching j_1 before reaching j_2. Then, the quantity

$$1 - s(a; j_2, j_1) - s(a; j_1, j_2)$$

represents the probability that $x^{(a)}(t)$ stays forever in J. This probability is 0 as the following theorem shows.

THEOREM 5.5.1. (A) *As a moves from j_1 to j_2, $s(a; j_2, j_1)$ increases (strictly) from 0 to 1 continuously.*

(B) $s(a; j_2, j_1) + s(a; j_1, j_2) = 1$.

PROOF. We will prove the assertions in five steps.

(i) If $j_1 \leq a < b \leq j_2$, then

$$s(a; j_2, j_1) = s(a; b, j_1)s(b; j_2, j_1) \tag{5.5.1}$$

holds. By definition $s(a; j_2, j_1)$ is the probability that a process starting from a reaches j_2 before it reaches j_1, but for this to happen the process must reach b before reaching j_1 and then it must start from b to reach j_2 before reaching j_1. Let τ be the time $x^{(a)}(t)$ reaches b before it reaches j_1 (if this does not occur, set $\tau = \infty$). Then τ is a Markov time so that we can apply the strong Markov property to obtain the desired result (5.5.1).

(ii) If $j_1 < a < j_2$, then

$$0 < s(a; j_2, j_1) < 1. \tag{5.5.2}$$

In order to prove this assertion, we first show that if $s(a; j_2, j_1) = 0$ for some a with $j_1 < a \leq j_2$, we get a contradiction. So, suppose there is such an a, and let B be the set of all b with $a < b \leq j_2$ for which $s(a; b, j_1) = 0$. By assumption, $j_2 \in B$. Let b_0 be the infimum of the set B. Take a point a' such that $j_1 < a' < b_0$, and let $b \in B$. Then, if $a' \leq a$, step (i) implies that

$$s(a'; b, j_1) = s(a'; a, j_1)s(a; b, j_1) = 0.$$

On the other hand, if $a < a' < b_0$, then

$$0 = s(a; b, j_1) = s(a; a', j_1)s(a'; b, j_1).$$

But since $a' < b_0$, we have $s(a; a', j_1) > 0$, so we can conclude that $s(a'; b, j_1) = 0$. Therefore, for all a' with $j_1 < a' < b_0$, we have $s(a'; b, j_1) = 0$ if $b \in B$. But this means that we have

$$P\{x^{(a')}(t) \leq b \quad \text{for } t \text{ in } 0 \leq t < \tau_J^{(a')}\} = 1.$$

By letting b approach b_0 within the set B, we obtain

$$P\{x^{(a')}(t) \leq b_0 \quad \text{for } t \text{ in } 0 \leq t < \tau_J^{(a')}\} = 1 \quad (j_1 < a' < b_0).$$

From assertion (ii) of the preceding section, it follows that b_0 must be a left shunt (although assertion (ii) of the preceding section refers to a right shunt, the corresponding condition, which is satisfied here, characterizes a left shunt). This contradicts our assumption that every point of J is a regular point. Thus, we conclude that $s(a; j_2, j_1) > 0$. Similarly, we have $s(a; j_1, j_2) > 0$, which in turn implies that

$$s(a; j_2, j_1) \leq 1 - s(a; j_1, j_2) < 1.$$

(iii) Combining (i) and (ii), we obtain

$$s(a; j_2, j_1) < s(b; j_2, j_1) \quad \text{when } j_1 \leq a < b \leq j_2.$$

Thus the proof of assertion (A) except that of the continuity is complete.

(iv) Let us now prove assertion (B). The quantity $\alpha = 1 - s(a; j_2, j_1) - s(a; j_1, j_2)$ is the probability for $x^{(a)}(t)$ to stay within J forever. Let us show that $\alpha = 0$. Let us call an open set U **dissipative** if for any $a \in U$, the probability for $x^{(a)}(t)$ to stay within U forever is 0. If U is dissipative and $U \supset V$, then so is V. Now, if $a \in J$, a is not a trap, so that $E(\tau_U^{(a)}) < \infty$ for a sufficiently small neighborhood U of a by assertion (xi) of §4.13. This implies that $P(\tau_U^{(a)} < \infty) = 1$, and therefore, U is dissipative. Next, let us show that if two open sub-intervals $U_1 = (u_1, v_1)$ and $U_2 = (u_2, v_2)$ of I, where $u_1 < u_2 < v_1 < v_2$, are dissipative, then so is their union $U = (u_1, v_2)$. So, let $a \in U_1 = (u_1, v_1)$. Then among the courses that $x^{(a)}(t)$ would follow, all of the following would lead it outside of U:

$$a \to u_1, \quad a \to v_1 \to v_2, \quad a \to v_1 \to u_2 \to u_1,$$
$$a \to v_1 \to u_2 \to v_1 \to v_2, \quad \cdots,$$

while the only possible course for $x^{(a)}(t)$ to follow which will keep it within U permanently is

$$a \to v_1 \to u_2 \to v_1 \to u_2 \to v_1 \to u_2 \to \cdots.$$

(Since both U_1 and U_2 are dissipative, there are no other possible courses for $x^{(a)}(t)$ to follow.) Therefore, the probability for $x^{(a)}(t)$ to stay in U permanently is given by the infinite product

$$s(a; v_1, u_1)s(v_1; u_2, v_2)s(u_2; v_1, u_1)s(v_1; u_2, v_2)s(u_2; v_1, u_1)\cdots$$

and since $s(v_1; u_2, v_2) < 1$, this infinite product is 0. When $a \in [v_1, v_2)$, the same argument shows that the same result follows, and therefore, U is dissipative.

By virtue of the Borel-Lebesgue covering theorem, we can cover $\overline{J}$ by a finite number of dissipative intervals. Then, by the result obtained above, $\overline{J}$ can be covered by a single dissipative open interval so that J itself must be dissipative. This implies that $\alpha = 0$ as desired.

(v) Finally, let us prove the continuity assertion of (A), which was left out in the argument above. First, let us show that

$$\lim_{b \uparrow j_2} s(a; b, j_1) = s(a; j_2, j_1) \tag{5.5.3}$$

holds. By the result of step (iv) above, we know that $\tau_J^{(a)}$ is finite with probability 1. If $x^{(a)}(\tau_J^{(a)}) = j_1$, then $x^{(a)}(t)$ must attain its maximum value for some t in $0 \le t \le \tau_J^{(a)}$, and this maximum is smaller than j_2. Therefore, if we choose b sufficiently close to (but smaller than) j_2, then $x^{(a)}(t)$ reaches j_1 before it reaches b. Consequently, if we take a sequence $b_n \uparrow j_2$, then if for any b_n, $x^{(a)}(t)$ reaches b_n before it reaches j_1, then it must be the case that $x^{(a)}(t)$ reaches j_2 before j_1. This means that

$$\lim_{n \to \infty} s(a; b_n, j_1) = s(a; j_2, j_1).$$

Furthermore, $s(a; b, j_1)$ decreases as b increases, for if $b < b'$, then $s(a; b', j_1) = s(a; b, j_1)s(b; b', j_1) \le s(a; b, j_1)$. Hence the sequential limit above must be the same as the limit in (5.5.3), i.e., (5.5.3) is proved.

Now, from step (i) and (5.5.3), we get

$$s(b; j_2, j_1) = \frac{s(a; j_2, j_1)}{s(a; b, j_1)} \to 1 \quad (b \uparrow j_2),$$

which implies that $s(b; j_2, j_1)$ is continuous at j_2. Similarly, we can show that $s(b; j_1, j_2)$ is continuous at $b = j_1$. Taking these facts into account, we obtain that

$$s(a; j_2, j_1) = s(a; b, j_1)s(b; j_2, j_1) \to s(b; j_2, j_1) \quad \text{as } a \uparrow b.$$

Similarly, when $a \downarrow b$, we have $s(a; j_1, j_2) \to s(b; j_1, j_2)$ so that

$$s(a; j_2, j_1) = 1 - s(a; j_1, j_2) \to 1 - s(b; j_1, j_2) = s(b; j_2, j_1) \quad \text{as } a \downarrow b.$$

Thus, we can conclude that $s(a; j_1, j_2)$ is continuous in a. □

THEOREM 5.5.2. *If $(j_1, j_2) \subset (k_1, k_2)$, then for $a \in [j_1, j_2]$ we have*

$$s(a; k_2, k_1) = s(a; j_1, j_2)s(j_1; k_2, k_1) + s(a; j_2, j_1)s(j_2; k_2, k_1).$$

PROOF. The result follows immediately if we apply the strong Markov property to a Markov time $\tau^{(a)}_{(j_1,j_2)}$. □

Substituting $s(a; j_1, j_2) = 1 - s(a; j_2, j_1)$ into the identity obtained in the theorem above, we get

$$s(a; k_2, k_1) = \alpha s(a; j_2, j_1) + \beta,$$

where α and β are constants depending on the two intervals (j_1, j_2) and (k_1, k_2). Using this fact, we obtain the following fundamental theorem.

THEOREM 5.5.3. *There exists a continuous, strictly increasing function $s(a)$ defined on $I = (i_1, i_2)$ unique up to a linear dependence such that for any $J = (j_1, j_2)$ with $\overline{J} \subset I$,*

$$s(a; j_2, j_1) = \frac{s(a) - s(j_1)}{s(j_2) - s(j_1)}$$

holds.

PROOF. To define one such function $s(a)$, fix a sub-interval $J^0 = (j_1^0, j_2^0)$ such that $\overline{J^0} \subset I$, and an arbitrary interval $J = (j_1, j_2)$ containing a and satisfying $I \supset \overline{J} \supset \overline{J^0}$, and set

$$s(a) = \alpha s(a; j_2, j_1) + \beta,$$

where α and β are chosen so that

$$s(j_2^0) = 1, \quad s(j_1^0) - 0.$$

If we take, instead of J, a bigger interval $J' = (j_1', j_2')$ and define $s'(a)$ in the same way as $s(a)$ was defined, by using J' in place of J, then by virtue of the preceding theorem, there exists a linear dependence between $s(b; j_2', j_1')$ and $s(b; j_2, j_1)$ if $b \in (j_1, j_2)$. Consequently, $s'(b)$ and $s(b)$ are also linearly dependent on each other. Furthermore, by definition, we have $s'(j_2^0) = s(j_2^0) = 1$, $s'(j_1^0) = s(j_1^0) = 0$. Hence, we actually have $s'(b) \equiv s(b)$ $(b \in (j_1, j_2))$. Taking $b = a$, we have in particular that $s'(a) = s(a)$, and this means that the value of $s(a)$ is determined independently of the choice of the sub-interval J. Also, it is clear from the definition of $s(a)$ that $s(a; j_2, j_1)$ is linearly dependent on $s(a)$, and since we furthermore have $s(j_2; j_2, j_1) = 1$ and $s(j_1, j_2, j_1) = 0$, we can conclude that the identity stated in the theorem must hold.

Conversely, if there are two functions $s_1(a)$ and $s_2(a)$ satisfying the conditions of the theorem, then in an arbitrary sub-interval J with $\overline{J} \subset I$, these two functions depend linearly on each other. Since the coefficients of this linear dependence are determined by values of s_1 and s_2 at two arbitrarily chosen a's, it is clear that these coefficients are determined independently of the choice of the sub-interval J. □

The function $s(a)$ determined in the preceding theorem is called the **canonical scale** in the interval I. It has an intrinsic probabilistic meaning, and is determined independently of the choice of coordinates in I (i.e., the choice of a particular homeomorphism between I and an interval on the real line). This canonical scale as well as the canonical measure, to be discussed in the next section, were first introduced by W. Feller.

EXAMPLE 5.5.1. Let $x^{(a)}(t)$ be the Wiener process on $R = R^1 \cup \{\infty\}$. If we let $I = R^1$, then I consists only of regular points. Let us show that $s(x) \equiv \alpha x + \beta$ holds for this process. Due to the symmetry property (in regard to left and right) of the Wiener process, it is easy to see that

$$s\left(\frac{j_1 + j_2}{2}; j_2, j_1\right) = \frac{1}{2},$$

from which it follows, by Theorem 5.5.3, that

$$s\left(\frac{j_1 + j_2}{2}\right) = \frac{1}{2}(s(j_1) + s(j_2)).$$

Since $s(x)$ is continuous, we can conclude from this that it must have the form $s(x) \equiv \alpha x + \beta$.

5.6. Feller's Canonical Measure

Let us first outline the procedure for defining the canonical measure. Let I be a regular interval as in the preceding section. Take a sub-interval J satisfying $\overline{J} \subset I$, and set

$$p_J(a) = E(\tau_J^{(a)}), \quad q_J(a) = -p_J(a).$$

We will show that $q_J(a)$ is convex with respect to $s(a)$, and that

$$m_J(a) = \frac{dq_J(a)}{ds(a)}$$

is an increasing function of a. $m_J(a)$ thus defined, however, is not necessarily continuous. If we take another interval J' such that $J \subset J'$, $\overline{J'} \subset I$, then for $a \in J$, we would have

$$m_J(a) = m_{J'}(a) + \text{const}.$$

Hence by normalizing $m_J(a)$ in such a way that $m_J(a_0) = 0$ for some fixed (but arbitrarily chosen) point $a_0 \in I$, we would have $m_J(a) = m_{J'}(a)$, so by setting this value as $m(a)$, we obtain an increasing function defined on the interval I. The measure dm obtained from this m is the **canonical measure** introduced by Feller. When s is fixed, then m is determined up to an additive constant, and if $s' = \alpha s + \beta$, then $m' = \alpha^{-1} m + \beta'$ must hold. Let us fill in the details by proving the following three theorems.

THEOREM 5.6.1. $p_J(a) < \infty$.

PROOF. Just as in the proof of the fact that J is dissipative, which was given in the preceding section, we take $J_1 = (u_1, v_1)$, $J_2 = (u_2, v_2)$ with $u_1 < u_2 < v_1 < v_2$, and show that the assumption $p_{J_1}(a), p_{J_2}(a) < \infty$ would imply $p_J(a) < \infty$, where $J = (u_1, v_2)$. As we know that $x^{(a)}(t)$ would leave the interval J with probability 1, we compute $p_J(a) = E(\tau_J^{(a)})$ by considering the courses the process $x^{(a)}(t)$ has to take in order to go out of J. If $a \in J_1$, then by using the strong Markov property, we get

$$\begin{aligned} p_J(a) &= p_{J_1}(a) + s(a; v_1, u_1)p_{J_2}(v_1) + s(a; v_1, u_1)s(v_1; u_2, v_2)p_{J_1}(u_2) \\ &\quad + s(a; v_1, u_1)s(v_1; u_2, v_2)s(u_2; v_1, u_1)p_{J_2}(v_1) + \cdots . \end{aligned}$$

Since $s(v_1; u_2, v_2) < 1$, this infinite series is majorized by a convergent geometric series and hence is convergent. (If $a \in J_2$, a similar argument also shows that $p_J(a) < \infty$.) □

THEOREM 5.6.2. *$q_J(a)$ is convex with respect to $s(a)$, and hence is a continuous function of a.*

PROOF. Using the strong Markov property, we get for $j_1 < a_1 < a < a_2 < j_2$,

$$p_J(a) = p_{(a_1,a_2)}(a) + s(a; a_1, a_2)p_J(a_1) + s(a; a_2, a_1)p_J(a_2).$$

(This fact was proved in a more general form in the proof of Dynkin's Theorem in §4.14.) By expressing $s(a; a_1, a_2)$ and $s(a; a_2, a_1)$ in the identity above by means of $s(a)$, we obtain

$$\begin{aligned} p_J(a) &= p_{(a_1,a_2)}(a) + \frac{s(a_2) - s(a)}{s(a_2) - s(a_1)}p_J(a_1) + \frac{s(a) - s(a_1)}{s(a_2) - s(a_1)}p_J(a_2) \\ &\geq \frac{s(a_2) - s(a)}{s(a_2) - s(a_1)}p_J(a_1) + \frac{s(a) - s(a_1)}{s(a_2) - s(a_1)}p_J(a_2), \end{aligned}$$

which shows that $p_J(a)$ is a concave function with respect to $s(a)$, and therefore, $q_J(a) = -p_J(a)$ is convex with respect to $s(a)$. □

By virtue of this theorem, $m_J(a) = \frac{dq_J(a)}{ds(a)}$ can now be determined and becomes an increasing function.

THEOREM 5.6.3. *If $J \subset J'$, then $m_J(a) \equiv m_{J'}(a) + \text{const.}$ $(a \in J)$ holds.*

PROOF. Just as for the proof of the preceding theorem, we have for $a \in J$,

$$p_{J'}(a) = p_J(a) + \frac{s(j_2) - s(a)}{s(j_2) - s(j_1)}p_J(j_1) + \frac{s(a) - s(j_1)}{s(j_2) - s(j_1)}p_J(j_2).$$

If we differentiate both sides of the identity above with respect to s, we obtain constants from the last two terms of the right-hand side, and we end up with

$$m_{J'}(a) = m_J(a) + \text{const.}$$ □

5.7. Feller's Canonical Form

Let I be a regular interval for $x^{(a)}(t)$, and let $s(x)$ and $dm(x)$ be the canonical scale and the canonical measure, respectively. Denote by A_I the local generator of $x^{(a)}(t)$ in I. The purpose of this section is to prove **Feller's canonical form**:

(5.7.1) $$A_I = (D_m D_s^+)_I$$

for A_I.

Let us begin by giving the precise definition of $D_m D_s^+$. For a fixed x, we set, as in the definition of derivatives in the ordinary sense,

$$D_s^+ f(x) = \lim_{\epsilon \downarrow 0} \frac{f(x+\epsilon) - f(x)}{s(x+\epsilon) - s(x)},$$

$$D_m D_s^+ f(x) = \lim_{\epsilon, \epsilon' \downarrow 0} \frac{D_s^+ f(x+\epsilon) - D_s^+ f(x-\epsilon')}{m(x+\epsilon) - m(x-\epsilon')}.$$

In order to make these definitions meaningful, it is only necessary to require that f is defined in some neighborhood of the point x. Now, suppose there is a function defined in some open interval I. If for such an f, $D_m D_s^+ f(x)$ as defined above can be determined for all $x \in I$, and furthermore, is continuous in x, then we say that

$$f \in \mathfrak{D}((D_m D_s^+)_I) \quad \text{and} \quad (D_m D_s^+)_I f(x) = D_m D_s^+ f(x). \tag{5.7.2}$$

In order for f to satisfy (5.7.2), f must of course be continuous. The assertion (5.7.1) means that the two operators must coincide with each other both in their domains of definition and in their values.

LEMMA 5.7.1. *$s \in \mathfrak{D}(A_I)$ and $A_I s(x) = 0$.*

PROOF. Since 0 is a continuous function, it suffices to show that $A_b s = 0$ for every $b \in I$. The definition of A_b is as it was given in Definition 5.3.2. Now, take a neighborhood $J = (j_1, j_2)$ of b in such a way that $\overline{J} \subset I$. If we set $s_J(x) = s(x; j_2, j_1)$, then since we can write $s(x) = \alpha s_J(x) + \beta$, we may prove $A_b s_J = 0$ instead of $A_b s = 0$. By using the Markov property, we obtain

$$\begin{aligned} s_J(b) &= \int_J P_J(t, b, dy) s_J(y) + P(\tau_J^{(b)} \le t, x^{(b)}(\tau_J^{(b)}) = j_2) \\ &= \int_J P_J(t, b, dy) s_J(y) + o(t). \end{aligned}$$

Therefore, we have

$$\frac{1}{t}\left\{ \int_J P_J(t, b, dy) s_J(y) - s_J(b) \right\} = o(1),$$

i.e., $A_b s_J = 0$. □

LEMMA 5.7.2. *Set $q(x) = \int_{x_0}^x m(y) ds(y)$, where x_0 is an arbitrarily chosen fixed point in I, and x varies over I. Then $q \in \mathfrak{D}(A_I)$ and*

$$A_I q(x) = 1.$$

PROOF. As for the preceding lemma, it suffices to prove that $A_b q = 1$ for an arbitrary $b \in I$. Consider a neighborhood J of b such that $\overline{J} \subset I$. Then, since we know that for $x \in J$

$$\frac{dq_J(x)}{ds(x)} = m(x) + \text{const.},$$

we have $q(x) = q_J(x) + \alpha s(x) + \beta$. As we already have $A_b s = 0$ from the preceding lemma, it is enough to show that $A_b q_J = 1$, that is, $A_b p_J = -1$. Using the Markov property, we get

$$\begin{aligned} p_J(b) &= \int_J P_J(t,b,dy)(p_J(y)+t) + E(\tau_J^{(b)}; \tau_J^{(b)} < t) \\ &= \int_J P_J(t,b,dy)p_J(y) + P_J(t,b,J)\cdot t + t\cdot o(t) \\ &= \int_J P_J(t,b,dy)p_J(y) + (1-o(t))\cdot t + t\cdot o(t), \end{aligned}$$

from which we obtain

$$\frac{1}{t}\left\{\int_J P_J(t,b,dy)p_J(y) - p_J(b)\right\} = -1 + o(t) \to -1. \qquad \square$$

LEMMA 5.7.3. *If* $f \in \mathfrak{D}(A_I)$ *and* $A_I f(b) > 0$, *then in some neighborhood* J *of* b, $f(x)$ *is convex with respect to* $s(x)$.

PROOF. Since $A_I f$ is continuous and $A_I f(b) > 0$, $A_I f(x) > 0$ in some neighborhood J of b. Choose two arbitrary points a_1, a_2 ($a_1 < a_2$) in J, and let α and β satisfy

$$f(a_i) = \alpha s(a_i) + \beta, \quad i = 1,2.$$

Then, in order to prove the assertion of the lemma, it suffices to prove that for $a_1 \le x \le a_2$,

$$f(x) \le \alpha s(x) + \beta$$

holds. So, set $g(x) \equiv f(x) - \alpha s(x) - \beta$. Then, since $g(x)$ is a continuous function, it must assume the maximum value $g(a_0)$ over the closed interval $[a_1, a_2]$ at some point a_0 in this interval. If $a_0 = a_1$ or $a_0 = a_2$, then, since $g(a_0) = g(a_i) = f(a_i) - \alpha s(a_i) - \beta = 0$, we have $g(x) \le 0$ for $x \in [a_1, a_2]$, i.e., the inequality $f(x) \le \alpha s(x) + \beta$ holds for $x \in [a_1, a_2]$. On the other hand, if a_0 is an interior point of (a_1, a_2), then we must have $A_I g(a_0) \le 0$ (cf. §5.3). But this means that

$$A_I f(a_0) \le \alpha A_I s(a_0) + \beta A_I 1 = 0,$$

which contradicts the assumption that $A_I f(x) > 0$ ($x \in J$). Therefore, the point a_0 must coincide with either a_1 or a_2, and we have the desired convexity of $f(x)$ with respect to $s(x)$ in J. $\square$

THEOREM 5.7.1. *If* $f \in \mathfrak{D}(A_I)$, *then* $f \in \mathfrak{D}((D_m D_s^+)_I)$, *and in* I

$$A_I f = (D_m D_s^+)_I f, \quad \text{i.e.,} \quad A_x f = D_m D_s^+ f(x), \quad x \in I,$$

holds.

PROOF. Let $f \in \mathfrak{D}(A_I)$ and $\alpha = A_I f(b)$. If we set

$$g(x) = f(x) - (\alpha - \delta) q(x),$$

then we have

$$A_I g(b) = \alpha - (\alpha - \delta)\cdot 1 = \delta > 0$$

so that $g(x)$ is convex with respect to $s(x)$ in some neighborhood J of b, and $D_s^+ g(x)$ is an increasing function on J. Consequently, if we take points b_1, b_2 in J in such a way that $b_2 > b > b_1$, then we have

$$D_s^+ g(b_2) > D_s^+ g(b_1),$$

from which it follows that

$$D_s^+ f(b_2) - D_s^+ f(b_1) > (\alpha - \delta)(m(b_2) - m(b_1)).$$

Since $m(x)$ is also increasing in J, we get

$$\frac{D_s^+ f(b_2) - D_s^+ f(b_1)}{m(b_2) - m(b_1)} > \alpha - \delta.$$

Since $\delta > 0$ can be taken as small as we please, and the choice of b_1, b_2 is arbitrary as long as $b_2 > b > b_1$, we can finally conclude from the above that

$$\varliminf_{\epsilon,\epsilon' \downarrow 0} \frac{D_s^+ f(b+\epsilon) - D_s^+ f(b-\epsilon')}{m(b+\epsilon) - m(b-\epsilon')} \geq \alpha.$$

Similarly, by considering $\alpha+\delta$ instead of $\alpha-\delta$, and arguing in the same way, we can show that $\varlimsup_{\epsilon,\epsilon' \downarrow 0} \leq \alpha$ holds, and therefore, we have $D_m D_s^+ f(b) = \alpha = A_I f(b)$. Since $A_I f(b)$ is continuous in b, we finally conclude that $(D_m D_s^+)_I f(b) = A_I f(b)$. □

We have the following converse to the preceding theorem.

THEOREM 5.7.2. *If $f \in \mathfrak{D}((D_m D_s^+)_I)$, then $f \in \mathfrak{D}(A_I)$ and $(D_m D_s^+)_I f = A_I f$ holds in I.*

PROOF. Suppose that every point of R is a one-dimensional diffusive point for $x^{(a)}(t)$. Let I be a regular interval for $x^{(a)}(t)$. Let us first define the operator L_b in the following way:

$$L_b f = \begin{cases} D_m D_s^+ f(b), & \text{if} \quad b \in I, \\ A_b f, & \text{if} \quad b \notin I. \end{cases}$$

We then define L by setting

$$\mathfrak{D}(L) = \{f / f \in \mathfrak{D}(L_b) \text{ for all } b \in R \text{ and } L_b f \text{ is continuous in } b \in R\},$$
$$Lf(b) \equiv L_b f.$$

Suppose now $f \in \mathfrak{D}(A)$. Then $f \in \mathfrak{D}(A_I)$, so that by the preceding theorem $f \in \mathfrak{D}((D_m D_s^+)_I)$. Therefore, if $b \in I$, then $f \in \mathfrak{D}(L_b)$ and $L_b f = D_m D_s^+ f(b) = A_b f = Af(b)$. On the other hand, if $b \notin I$, then since $f \in \mathfrak{D}(A)$, we have $f \in \mathfrak{D}(A_b)$ so that $f \in \mathfrak{D}(L_b)$ and $L_b f = A_b f = Af(b)$. So, in all cases, $f \in \mathfrak{D}(L_b)$ and $L_b f = A_b f = Af(b)$. As $Af(b)$ is a continuous function of b, it follows that $f \in \mathfrak{D}(L)$ and $Af = Lf$. This means that $L \supset A$ holds.

Next, we show that $(\lambda - L)^{-1}$ ($\lambda > 0$) exists. For this purpose, it suffices to show that $\lambda u = Lu$ would imply $u \equiv 0$. So, assume that $\lambda u = Lu$. Let $u(a)$ be the minimum value for u. If $a \in I$, then $Lu(x) = D_m D_s^+ u(x)$ holds in a neighborhood of a. If $u(a) < 0$, then $Lu(a) < 0$ so that $D_m D_s^+ u(x) < 0$ holds in a neighborhood of a. But this would imply that $D_s^+ u(x)$ is decreasing, and hence $-u(x)$ becomes convex with respect to $s(x)$ in this neighborhood of a. But this would mean that $-u(x)$ cannot take its maximum at a, or $u(x)$ cannot take its minimum at a, which contradicts the assumption. Hence if u takes its minimum at a and if $a \in I$, then we must have $u(a) \geq 0$. If $a \notin I$, then $Lu(a) = A_a u$ and since $u(a)$ is minimum, we must have $A_a u \geq 0$ by the definition of A_a (cf. §5.3). Hence, $u(a) = \lambda^{-1} Lu(a) \geq 0$. Thus, we have shown that $u(a) \geq 0$ must hold if $u(a)$ is minimum, in all cases, and thus, $u(x) \geq 0$ for all x. If we consider $-u$ instead of u, which also satisfies $\lambda(-u) = L(-u)$, we get by the same argument that $-u(x) \geq 0$ for all x, and thus we conclude that $u \equiv 0$ as desired.

Now, from $L \supset A$, we have $(\lambda - L)^{-1} \supset (\lambda - A)^{-1} = R_\lambda$. But since the domain of definition of R_λ is all of $\boldsymbol{C}$, the space of all continuous functions on R, we conclude that $(\lambda - L)^{-1} = (\lambda - A)^{-1}$, from which it follows that $L = A$.

In order to complete the proof of the theorem, we have to show that if $f \in \mathfrak{D}((D_m D_s^+)_I)$ and if b is an arbitrary point of I, then $A_b f$ can be determined, and that $A_b f$ is continuous in b. It suffices to show that $A_b f$ is continuous in an arbitrary sub-interval $K = (k_1, k_2)$ with $\overline{K} \subset I$. We may also assume that the end points of K are continuity points of the function m. Let us take an interval $J = (j_1, j_2)$ satisfying $\overline{K} \subset J \subset \overline{J} \subset I$, and suppose that the end points of J also are continuity points of m. Suppose we can construct a function g having the following properties:

$$g(b) = \begin{cases} f(b), & \text{if} \quad b \in K, \\ 0, & \text{if} \quad b \notin J, \end{cases} \qquad \text{and } D_m D_s^+ g(b) \text{ is continuous in } b \in I.$$

Then g belongs to the domain of definition of L and hence of A, and $A_b g = Ag(b)$ is continuous in $b \in J$. As $f = g$ holds in K, we have $A_b f = A_b g$ by the local property of A_b, and hence we can conclude that $A_b f$ is continuous in $b \in K$ as desired. Thus it remains only to construct such a function g. It is clear that it is enough to construct the function $h = Lg$ instead of g itself. Such an h must satisfy

$$h(b) = \begin{cases} D_m D_s^+ f(b), & \text{if} \quad b \in K, \\ 0, & \text{if} \quad b \notin J, \end{cases}$$

so it is enough to define h on the union $[j_1, k_1] \cup [k_2, j_2]$ of two disjoint closed intervals. Let us consider the case of $[j_1, k_1]$. As the requirements for h are "the continuity of $D_s^+ h$ at the continuity points of m" and "the continuity of $D_m D_s^+ h$", the necessary and sufficient conditions for h on the interval $[j_1, k_1]$ are given by

$$\begin{aligned}
&\ell_1(h) \equiv h(j_1) = 0, \\
&\ell_2(h) \equiv h(k_1) = D_m D_s^+ f(k_1), \\
&\ell_3(h) \equiv \int_{j_1}^{k_1} h(x) dm(x) = D_s^+ f(k_1), \\
&\ell_4(h) \equiv \int_{j_1}^{k_1} \int_{j_1}^{y} h(x) dm(x) ds(y) \equiv \int_{j_1}^{k_1} h(x)(s(k_1) - s(x)) dm(x) = f(k_1).
\end{aligned}$$

It is clear that the linear functionals $\ell_1, \ell_2, \ell_3, \ell_4$ defined above on the Banach space $C[j_1, k_1]$ are linearly independent, and therefore, there exists a continuous function h satisfying these conditions. We can similarly show the existence of h satisfying necessary and sufficient conditions for the interval $[k_2, j_2]$, and thus the proof of this theorem is complete. □

5.8. Local Generators at Generalized Shunts

In the sequel, we will simply call an open set I of R an open interval if it is homeomorphic to an open interval (r_1, r_2) on the real line. When we say that a point a is an end point of an open interval I, it means that the set $\{a\} \cup I$ is homeomorphic to a half open interval $[r_1, r_2)$ or $(r_1, r_2]$. Obviously, the point a will be mapped onto r_1 (or r_2) under this homeomorphism.

Let a be an end point of an open interval I consisting only of diffusive points. If for a sufficiently small neighborhood U of a

$$P\{x^{(a)}(t) \in \{a\} \cup I \text{ is satisfied for every } t \in [0, \tau_U^{(a)})\} = 1 \tag{5.8.1}$$

holds, then a is called a **generalized shunt**. A generalized shunt becomes a shunt (cf. §5.4) only when it has a one-dimensional character. As we have distinguished left and right shunts before, depending on the orientation of the image of a neighborhood of such points under a particular class of homeomorphisms onto the corresponding interval on the real line, we will distinguish generalized shunts by calling a a **generalized right shunt** if we consider $\{a\} \cup I$ to be mapped homeomorphically onto $[r_1, r_2)$, and a **generalized left shunt** if mapped onto $(r_1, r_2]$. In what follows we consider only the case of generalized right shunts.

It is, of course, possible for a generalized right shunt to be a trap, but since such a case would be too simple, we will exclude this possibility. Then we can assume, in addition to (5.8.1), that for a sufficiently small neighborhood U of a,

$$E(\tau_U^{(b)}) < \infty, \quad \text{for } b \in U. \tag{5.8.2}$$

Let us represent points of $\{a\} \cup I$ by the corresponding points of $[r_1, r_2)$. Since the points sufficiently close to r_1 lie in the neighborhood U, if we set $r_2' = \sup\{\xi / [r_1, \xi) \subset U\}$, then it is clear that $r_2' > r_1$. If we set, for $r_1 < \xi < r_2'$,

$$p(\xi) = E(\tau_{[r_1,\xi)}^{(r_1)}), \tag{5.8.3}$$

then $0 \le p(\xi) < \infty$. Since $x^{(r_1)}(t)$ always lies in $[r_1, r_2)$ for $0 \le t < \tau_U^{(r_1)}$ and since $x^{(r_1)}(t)$ approaches r_2' as t approaches $\tau_U^{(r_1)}$, for each ξ in (r_1, r_2'), there exists at least one t in $[0, \tau_U^{(r_1)})$ such that $x^{(r_1)}(t) = \xi$. We denote by τ^* the smallest of such t's, and use the strong Markov property with respect to τ^* to obtain

$$p(\xi) + E(\tau_U^{(\xi)}) = E(\tau_U^{(r_1)}). \tag{5.8.4}$$

Since $[r_1, r_2')$ consists of one-dimensional diffusive points, we can show, as we did for the case of regular points (cf. Lemma 5.7.2), that $p_U(\xi) = E(\tau_U^{(\xi)})$ satisfies

$$A_\xi p_U = -1.$$

Therefore, we have

LEMMA 5.8.1. $A_\xi p = 1$.

LEMMA 5.8.2. *Suppose $f \in \mathfrak{D}(A_V)$, where V is some neighborhood of $a = r_1$, and f satisfies*

$$A_V f(r_1) > 0.$$

Then for a sufficiently small $\epsilon > 0$, we have

$$f(\xi) \ge f(r_1), \quad r_1 < \xi < r_1 + \epsilon.$$

PROOF. Since $A_V f(r_1) > 0$, we have for a sufficiently small $\epsilon > 0$,

$$A_V f(\xi) > 0, \quad r_1 < \xi < r_1 + \epsilon.$$

Therefore, $f(\xi)$ cannot attain a relative maximum in the interval $(r_1, r_1 + \epsilon)$. Thus, $f(\xi)$ must either be non-increasing or non-decreasing in some neighborhood $r_1 \le \xi < r_1 + \delta$ of $\xi = r_1$. But if it is non-increasing in $r_1 \le \xi < r_1 + \delta$, then we must have $A_{r_1} f \le 0$ so that $A_V f(r_1) \le 0$, but this contradicts the assumption of the lemma. □

THEOREM 5.8.1. *If $f \in \mathfrak{D}(A_V)$, then*

$$A_V f(r_1) = \lim_{\xi \downarrow r_1} \frac{f(\xi) - f(r_1)}{p(\xi)}.$$

PROOF. Set $\alpha = A_V f(r_1)$, and define $g(\xi) = f(\xi) - (\alpha - \epsilon)p(\xi)$, $r_1 < \xi < r_2$. Then $g \in \mathfrak{D}(A_V)$ and

$$A_V g(r_1) = \alpha - (\alpha - \epsilon)1 = \epsilon > 0.$$

Therefore, for a sufficiently small $\delta > 0$, we have

$$g(\xi) \geq g(r_1), \quad r_1 < \xi < r_1 + \delta,$$

from which it follows that

$$\frac{f(\xi) - f(r_1)}{p(\xi)} > \alpha - \epsilon, \quad r_1 < \xi < r_1 + \delta.$$

It is easy to deduce from this that we have

$$\underline{\lim}_{\xi \downarrow r_1} \frac{f(\xi) - f(r_1)}{p(\xi)} \geq \alpha.$$

Considering $\alpha + \epsilon$ in place of $\alpha - \epsilon$ in the argument above, and reasoning in the same way, we also obtain

$$\overline{\lim}_{\xi \downarrow r_1} \frac{f(\xi) - f(r_1)}{p(\xi)} \leq \alpha. \qquad \square$$

5.9. Distribution of the First Passage Time

Let $I = (r_1, r_2)$ be a regular interval for $x^{(a)}(t)$, and let s and dm be the canonical scale and the canonical measure, respectively. Take a sub-interval $J = (j_1, j_2)$ satisfying $\overline{J} \subset I$, and let a be an arbitrary point of J. Denote by $\tau_J^{(a)}$ the first passage time of J for $x^{(a)}(t)$. It is clear that $\tau_J^{(a)}$ has a finite mean, and hence it is finite with probability 1. The point $x^{(a)}(\tau_J^{(a)})$ is either j_1 or j_2, so we define

$$\tau_1^{(a)} = \tau_J^{(a)} \text{ and } \tau_2^{(a)} = \infty, \text{ if } x^{(a)}(\tau_J^{(a)}) = j_1,$$
$$\tau_2^{(a)} = \tau_J^{(a)} \text{ and } \tau_1^{(a)} = \infty, \text{ if } x^{(a)}(\tau_J^{(a)}) = j_2.$$

The condition $\tau_i^{(a)} < \infty$ is equivalent to $\tau_i^{(a)} = \tau_J^{(a)}$, and hence to $x^{(a)}(\tau_J^{(a)}) = j_i$. As we have shown in §5.7, we have

$$A_J = (D_m D_s^+)_J. \tag{5.9.1}$$

We also have $s_J(a) = s(a; j_2, j_1) = P(\tau_2^{(a)} < \infty)$, and

$$\begin{cases} A_J s_J(a) = 0, \\ s_J(j_1 + 0) = 0, \quad s_J(j_2 - 0) = 1, \end{cases} \tag{5.9.2}$$

which can be deduced easily from (5.9.1) and the fact that

$$s_J(a) = \frac{s(a) - s(j_1)}{s(j_2) - s(j_1)}.$$

Next, let us put $p_J(a) = E(\tau_J^{(a)})$. Then we have, as was shown in §5.7, that

$$A_J p_J(a) = -1 \tag{5.9.3}$$

holds. Furthermore, we have

$$p_J(j_1+0) = p_J(j_2-0) = 0. \tag{5.9.4}$$

In fact, since j_1 is a regular point, we have $p_U(x) < \epsilon$, $x \in U$, for a sufficiently small neighborhood U of j_1, where $p_U(x)$ denotes $E(\tau_U^{(x)})$. Choose a $\delta > 0$ so small that $K = [j_1, j_1+\delta] \subset U$ is satisfied. Then, as long as $x \in K$, we have

$$p_J(x) = p_K(x) + s(x;\xi,j_1)p_J(\xi) < \epsilon + s(x;\xi,j_1)p_J(\xi).$$

If we let $x \downarrow j_1$, then $s(x;\xi,j_1) \to 0$ so that we have

$$p_J(j_1+0) \leq \epsilon,$$

and letting $\epsilon \downarrow 0$, we obtain

$$p_J(j_1+0) = 0.$$

Similarly, we get

$$p_J(j_2-0) = 0.$$

THEOREM 5.9.1. *s_J and p_J satisfy, and are determined uniquely by, the properties* (5.9.2), *and* (5.9.3) *plus* (5.9.4), *respectively.*

PROOF. We have already shown that s_J and p_J satisfy these conditions. Conversely, since the equation $A_J u = 0$ has two linearly independent solutions as can be seen from (5.9.1), it is clear that the boundary conditions given above determine the solutions uniquely. □

Next for $i = 1, 2$, define

$$\varphi_i(dt, a) = P(\tau_i^{(a)} \in dt),$$

$$\hat{\varphi}_{i\lambda}(a) = \int_0^\infty e^{-\lambda t}\varphi_i(dt, a) = E(e^{-\lambda\tau_i^{(a)}}).$$

THEOREM 5.9.2. *$\hat{\varphi}_{1\lambda}$ is the unique solution of the boundary value problem*

$$A_J\hat{\varphi}_{1\lambda} = \lambda\hat{\varphi}_{1\lambda},\ \ \hat{\varphi}_{1\lambda}(j_1+0) = 1,\ \ \hat{\varphi}_{1\lambda}(j_2-0) = 0. \tag{5.9.5}$$

Similarly, $\hat{\varphi}_{2\lambda}$ is the unique solution of

$$A_J\hat{\varphi}_{2\lambda} = \lambda\hat{\varphi}_{2\lambda},\ \ \hat{\varphi}_{2\lambda}(j_2-0) = 1,\ \ \hat{\varphi}_{2\lambda}(j_1+0) = 0. \tag{5.9.6}$$

PROOF. We shall prove the assertion for $\hat{\varphi}_{1\lambda}$. As before, we set

$$P_J(t, a, dy) = P(x^{(a)}(t) \in dy, \tau_J^{(a)} > t).$$

Now,

$$\varphi_1(dt+s, a) = \int_J P_J(s, d, dy)\varphi_1(dt, y),$$

$$\int_0^\infty e^{-\lambda t}\varphi_1(dt+s, a) = \int_J P_J(s, a, dy)\int_0^\infty e^{-\lambda t}\varphi_1(dt, y),$$

$$e^{\lambda s}\int_s^\infty e^{-\lambda t}\varphi_1(dt, a) = \int_J P_J(s, a, dy)\hat{\varphi}_{1\lambda}(y).$$

But

$$\int_s^\infty e^{-\lambda t}\varphi_1(dt, a) = \hat{\varphi}_{1\lambda}(a) - \int_0^s e^{-\lambda t}\varphi_1(dt, a),$$

$$\int_0^s e^{-\lambda t}\varphi_1(dt, a) \leq \varphi_1([0, s], a) \leq P(\tau_J^{(a)} < s) = o(s).$$

Therefore, we have

$$e^{\lambda s}(\hat{\varphi}_{1\lambda}(a) + o(s)) = \int_J P_J(s, a, dy)\hat{\varphi}_{1\lambda}(y),$$

from which we obtain immediately

$$A_a\hat{\varphi}_{1\lambda} = \lambda\hat{\varphi}_{1\lambda}(a).$$

If we can show that the right-hand side of the identity above is continuous in a, then we would have proved

$$A_J\hat{\varphi}_{1\lambda} = \lambda\hat{\varphi}_{1\lambda} \text{ in } J.$$

So far, we defined $\hat{\varphi}_{1\lambda}(b)$ by starting with a sub-interval $J \subset I$. If we start with another sub-interval K to define the corresponding quantity, we denote it by $\hat{\varphi}_{1\lambda}(b; K)$. Then, by the same type of argument as those we have often used, using the strong Markov property, we obtain

$$\hat{\varphi}_{1\lambda}(b) = \hat{\varphi}_{1\lambda}(b; (a, j_2))\hat{\varphi}_{1\lambda}(a) \leq \hat{\varphi}_{1\lambda}(a) \quad \text{if} \quad a < b.$$

Therefore, $\hat{\varphi}_{1\lambda}$ is monotone decreasing. The continuity of $\hat{\varphi}_{1\lambda}$ would be established if we can show that for any $c \in J$ and $\epsilon > 0$ we can choose a and b sufficiently close to c in such a way that $a < c < b$, and for which

$$\hat{\varphi}_{1\lambda}(b; (a, j_2)) > 1 - \epsilon$$

is satisfied. Now,

$$\begin{aligned}
\hat{\varphi}_{1\lambda}(b; (a, j_2)) &= E\{e^{-\lambda\tau^{(b)}_{(a,j_2)}}; x^{(b)}(\tau^{(b)}_{(a,j_2)}) = a\} \\
&\geq e^{-\lambda t}P(\tau^{(b)}_{(a,j_2)} < t, x^{(b)}(\tau^{(b)}_{(a,j_2)}) = a) \\
&\geq e^{-\lambda t}\{P(\tau^{(b)}_{(a,k)} < t) - P(x^{(b)}(\tau^{(b)}_{(a,k)}) = k)\},
\end{aligned}$$

where k is a point lying between b and j_2. We also have

$$\begin{aligned}
P(\tau^{(b)}_{(a,k)} \geq t) &\leq \frac{1}{t}E(\tau^{(b)}_{(a,k)}), \\
P(x^{(b)}(\tau^{(b)}_{(a,k)}) = k) &= \frac{s(b) - s(a)}{s(k) - s(a)}.
\end{aligned}$$

For the given ϵ, let $\delta = \epsilon/3$ and $t = \delta/\lambda$. Choose a sufficiently small neighborhood U of c so that $E(\tau_U^{(b)}) < \delta^2/\lambda$. Then, choose points a, b, k in U in such a way that $a < c < b < k$ and $(s(b) - s(a))/(s(k) - s(a)) < \delta$ are satisfied. This is possible by choosing a and b sufficiently close to c. Since

$$E(\tau^{(b)}_{(a,k)}) \leq E(\tau_U^{(b)}) < \delta^2/\lambda$$

holds, we obtain

$$\begin{aligned}
\hat{\varphi}_{1\lambda}(b; (a, j_2)) &\geq e^{-\lambda t}(1 - 2\delta) \\
&> (1 - \delta)(1 - 2\delta) > 1 - \epsilon \quad \text{as} \quad \delta = \epsilon/3.
\end{aligned}$$

Thus the continuity of $\hat{\varphi}_{1\lambda}$ on J is proved.

Next, we prove that the boundary conditions $\hat{\varphi}_{1\lambda}(j_1 + 0) = 1$, $\hat{\varphi}_{1\lambda}(j_2 - 0) = 0$ are satisfied. As we have seen above, $\hat{\varphi}_{1\lambda}(a)$ decreases as a increases, and its values lie between 0 and 1. If we take points k_1 and k_2 in such a way that $k_1 < j_1$,

$a < k_2 < j_2$ are satisfied, then by the same argument as we used for $\hat{\varphi}_{1\lambda}(b; (a, j_2))$ in the proof above of the continuity, we obtain

$$\hat{\varphi}_{1\lambda}(a) \geq e^{-\lambda t}\{P(\tau^{(a)}_{(k_1,k_2)} < t) - P(x^{(a)}(\tau^{(a)}_{(k_1,k_2)}) = k_2)\},$$

from which it follows that $\hat{\varphi}_{1\lambda}(a) > 1 - \epsilon$ is satisfied if a is sufficiently close to j_1. This shows that $\hat{\varphi}_{1\lambda}(j_1 + 0) = 1$ is satisfied. The assertion $\hat{\varphi}_{2\lambda}(j_2 - 0) = 1$ can be established in a similar manner. Since we have

$$\hat{\varphi}_{1\lambda}(a) + \hat{\varphi}_{2\lambda}(a) = E(e^{-\lambda\tau_J^{(a)}}) \leq 1,$$

we obtain

$$\hat{\varphi}_{1\lambda}(j_2 - 0) \leq 1 - \hat{\varphi}_{2\lambda}(j_2 - 0) = 0, \text{ and hence } \hat{\varphi}_{1\lambda}(j_2 - 0) = 0.$$

Finally, in order to prove the uniqueness, it suffices to show that if $A_J u = \lambda u$ and $u(j_1 + 0) = u(j_2 - 0) = 0$ are satisfied, then $u \equiv 0$ must hold. If $u > 0$ is possible, then u takes a positive maximum value at an interior point a of J. Then we must have $A_J u(a) \leq 0$ and hence $u(a) = \lambda^{-1} A_J u(a) \leq 0$, which is a contradiction. Thus, we must have $u \leq 0$. Considering $-u$ in place of u, we obtain $u \geq 0$, and therefore, we must have $u \equiv 0$. □

THEOREM 5.9.3. *If we set* $\hat{\varphi}_\lambda(a) = E(e^{-\lambda\tau_J^{(a)}})$, *then it is the unique solution of the boundary value problem*

$$A_J\hat{\varphi}_\lambda = \lambda\hat{\varphi}_\lambda, \quad \hat{\varphi}_\lambda(j_1 + 0) = \hat{\varphi}_\lambda(j_2 - 0) = 1.$$

PROOF. Since

$$\hat{\varphi}_\lambda(a) = E(e^{-\lambda\tau_J^{(a)}}) = \hat{\varphi}_{1\lambda}(a) + \hat{\varphi}_{2\lambda}(a),$$

where $\hat{\varphi}_{i\lambda}(a)$, $i = 1, 2$, are the functions introduced in the preceding theorem, we have

$$A_J\hat{\varphi}_\lambda = A_J\hat{\varphi}_{1\lambda} + A_J\hat{\varphi}_{2\lambda} = \lambda\hat{\varphi}_{1\lambda} + \lambda\hat{\varphi}_{2\lambda} = \lambda\hat{\varphi}_\lambda.$$

Furthermore, we also have

$$\hat{\varphi}_\lambda(j_1 + 0) = \hat{\varphi}_{1\lambda}(j_1 + 0) + \hat{\varphi}_{2\lambda}(j_1 + 0) = 1,$$
$$\hat{\varphi}_\lambda(j_2 - 0) = \hat{\varphi}_{1\lambda}(j_2 - 0) + \hat{\varphi}_{2\lambda}(j_2 - 0) = 1.$$

Therefore, $\hat{\varphi}_\lambda$ satisfies the equation and the boundary condition of the theorem. The uniqueness of the solution can be established in the same way as for the preceding theorem. □

5.10. Classical Diffusion Processes

Let R be either the real line R^1 or its closed sub-interval $[r_1, r_2]$. Suppose $x^{(a)}(t)$ is a diffusion process on R. Then by Ray's Theorem

(5.10.1) $$P\{|x^{(\xi)}(t) - \xi| > \epsilon\}/t \to 0 \quad (t \to 0)$$

holds. If we suppose that for some $\epsilon > 0$, both

(5.10.2) $$a(\xi) = \lim_{t\downarrow 0} \frac{E\{x^{(\xi)}(t) - \xi; |x^{(\xi)}(t) - \xi| < \epsilon\}}{t}$$

and

(5.10.3) $$b(\xi) = \lim_{t\downarrow 0} \frac{E\{(x^{(\xi)}(t) - \xi)^2; |x^{(\xi)}(t) - \xi| < \epsilon\}}{t}$$

exist, then, because of (5.10.1), they exist for any $\epsilon > 0$ and their values are independent of ϵ. When both $a(\xi)$ and $b(\xi)$ are continuous functions of ξ in (r_1, r_2), then we call $x^{(a)}(t)$ a classical diffusion process, or Kolmogorov's diffusion process on R $(= [r_1, r_2])$. In terms of the transition probability $P(t, a, E)$ of $x^{(a)}(t)$, $a(\xi)$ and $b(\xi)$ can be written as follows:

$$a(\xi) = \lim_{t \downarrow 0} \frac{1}{t} \int_{|y-\xi|<\epsilon} (y - \xi) P(t, \xi, dy), \tag{5.10.4}$$

$$b(\xi) = \lim_{t \downarrow 0} \frac{1}{t} \int_{|y-\xi|<\epsilon} (y - \xi)^2 P(t, \xi, dy). \tag{5.10.5}$$

$a(\xi)$ can take positive and negative values, while $b(\xi) \geq 0$.

For example, for the Wiener process, we have

$$a(\xi) = 0, \quad b(\xi) = 1. \tag{5.10.6}$$

THEOREM 5.10.1. *If f is twice continuously differentiable in $J = (j_1, j_2)$ $(r_1 \leq j_1 < j_2 \leq r_2)$, then $f \in \mathfrak{D}(A_J)$ and*

$$A_J f(\xi) = \left(a(\xi) \frac{d}{d\xi} + \frac{b(\xi)}{2} \frac{d^2}{d\xi^2} \right) f(\xi) \tag{5.10.7}$$

holds.

PROOF. Since the right-hand side of the equation above is continuous in ξ by hypothesis, in order to prove the theorem, it suffices to show that $A_\xi f$ equals the right-hand side. For $\xi \in J$ and $\delta > 0$, we can choose $\epsilon > 0$ sufficiently small so that whenever $|y - \xi| < \epsilon$, then

$$f(y) - f(\xi) = f'(\xi)(y - \xi) + \frac{f''(\xi) \pm \delta}{2} (y - \xi)^2.$$

Therefore, we have

$$\begin{aligned}
&\int_{|y-\xi|<\epsilon} f(y) P(t, \xi, dy) - f(\xi) = \int_{|y-\xi|<\epsilon} (f(y) - f(\xi)) P(t, \xi, dy) + o(t) \\
&= f'(\xi) \int_{|y-\xi|<\epsilon} (y - \xi) P(t, \xi, dy) + \frac{f''(\xi) \pm \delta}{2} \int_{|y-\xi|<\epsilon} (y - \xi)^2 P(t, \xi, dy) + o(t) \\
&= f'(\xi) a(\xi) \cdot t + \frac{f''(\xi) \pm \delta}{2} b(\xi) \cdot t + o(t).
\end{aligned}$$

It then follows that

$$\overline{\lim_{t \downarrow 0}} \frac{1}{t} \left\{ \int_{|y-\xi|<\epsilon} f(y) P(t, \xi, dy) - f(\xi) \right\} \leq a(\xi) f'(\xi) + \frac{b(\xi)}{2} f''(\xi) + \frac{b(\xi)}{2} \cdot \delta.$$

Since the left-hand side of this inequality is independent of ϵ by virtue of (5.10.1), we may take δ on the right-hand side as small as we please, and thus we can conclude that the limsup of the left-hand side above must be less than or equal to $a(\xi) f'(\xi) + b(\xi)/2 \cdot f''(\xi)$. Similarly, by taking liminf we can conclude that the liminf of the left-hand side above must be greater than or equal to $a(\xi) f'(\xi) + b(\xi)/2 \cdot f''(\xi)$. Thus, we see that $f \in \mathfrak{D}(A_\xi)$ and $A_\xi f = a(\xi) f'(\xi) + b(\xi)/2 \cdot f''(\xi)$. $\square$

THEOREM 5.10.2. *If $b(\xi) > 0$, then ξ is a regular point.*

PROOF. If ξ is a right shunt, then a function $f \in \mathfrak{D}(A_\xi)$, which is increasing in some neighborhood J of ξ, must satisfy $A_\xi f \geq 0$. But if we define f in some neighborhood of ξ by

$$f(x) = (x-\xi) - \beta(x-\xi)^2,$$

then we get

$$\left(af' + \frac{b}{2}f''\right)(\xi) = a(\xi) - \beta b(\xi),$$

which is negative if $\beta > a(\xi)/b(\xi)$ so that

$$A_\xi f = A_J f(\xi) = \left(af' + \frac{b}{2}f''\right)(\xi) < 0.$$

Since $f(x)$ is increasing in a neighborhood $|x-\xi| < 1/(2\beta)$ of ξ, this is a contradiction. Thus, ξ cannot be a right shunt. Similarly, we can show that ξ cannot be a left shunt either. Hence, we conclude that ξ must be a regular point. □

Now, for functions twice continuously differentiable in J, let us define the operator D_J by

$$D_J f(\xi) = a(\xi)f'(\xi) + \frac{b(\xi)}{2}f''(\xi). \tag{5.10.8}$$

THEOREM 5.10.3. *If $b(\xi) > 0$ in J, then*

$$A_J = D_J \tag{5.10.9}$$

holds.

PROOF. By Theorem 5.10.1, we have $A_J \supset D_J$. We rewrite D_J as follows:

$$\begin{aligned} D_J f &= \frac{b}{2}e^{-B}\left(e^B\frac{2a}{b}f' + e^B f''\right), \quad \text{where} \quad B = \int \frac{2a}{b}d\xi, \\ &= \frac{b}{2}e^{-B}(e^B f')' = \frac{1}{\frac{2}{b}e^B}\frac{d}{d\xi}\frac{1}{e^{-B}}\frac{d}{d\xi}f(\xi) \\ &= \frac{d}{dm_1}\frac{d}{ds_1}f(\xi), \quad \text{where} \quad s_1 = \int e^{-B}d\xi \text{ and } m_1 = \int \frac{2}{b}e^B d\xi. \end{aligned}$$

Since $A_J = (D_m D_s^+)_J$, the set of all solutions of $A_J u = 0$ must be of the form $\{\alpha + \beta s\}$, while the set of all solutions of $D_J u = 0$ is of the form $\{\alpha' + \beta' s_1\}$. From $A_J \supset D_J$, we have $\{\alpha + \beta s\} \supset \{\alpha' + \beta' s_1\}$, but as both of these two sets are two-dimensional linear spaces, they must coincide. Therefore, we can write $s = \alpha + \beta s_1$. Next, if we set

$$q = \int m_1 ds_1 = \int m_1 e^{-B} d\xi,$$

then $D_J q = 1$ and hence $A_J q = 1$, i.e., $D_m D_s^+ q = 1$ holds. Consequently,

$$q = \int m ds + \gamma s + \delta = \beta \cdot \int m ds_1 + \gamma' s_1 + \delta',$$

$$m_1 = \frac{dq}{ds_1} = \beta m + \gamma', \text{ and hence } m = \gamma'' + \frac{1}{\beta}m_1.$$

Therefore, we may suppose that $m = m_1$ and $s = s_1$. Now, if $f \in \mathfrak{D}(A_J)$, then $D_{m_1}D_{s_1}^+ f(\xi)$ is continuous in ξ. Since $2/b \cdot e^B$ is also continuous and strictly positive, the continuity of $D_{m_1}D_{s_1}^+ f(\xi) = \frac{1}{2/b\cdot e^B}(e^B f')'$ implies the continuity of $(e^B f')'$. Therefore, $e^B f'$ is continuously differentiable. Finally, since $(e^{-B})' = -e^{-B} \cdot 2a/b$,

e^{-B} is continuously differentiable, and hence so is $f' = e^{-B} \cdot (e^B f')$. Therefore, f must be twice continuously differentiable and hence $f \in \mathfrak{D}(D_J)$ is satisfied. This shows that $A_J \subset D_J$, and thus the theorem is proved. □

EXAMPLE 5.10.1. For the Wiener process, we have

$$D_J = \frac{1}{2}\frac{d^2}{dx^2}, \quad \mathfrak{D}(D_J) = \boldsymbol{C}_2(J).$$

By the preceding theorem, we must have the same for A_J. Let $J = (j_1, j_2)$ be a finite open interval, and let $s_J(x)$ be the probability for a process starting from x to reach the right end point j_2 before reaching j_1 (i.e., $s_J(x) = s(s; j_2, j_1)$). Then, as we have shown in the preceding section, s_J is the solution of

$$A_J u = 0, \quad u(j_2) = 1, \quad u(j_1) = 0.$$

As $A_J = D_J$, $A_J u = 0$ means $u'' = 0$, i.e., $u = \alpha + \beta x$. Therefore, we have

$$s_J(x) = \frac{x - j_1}{j_2 - j_1},$$

which shows that the function x is the canonical scale in this case.

We know also that $p_J(x) = E(\tau_J^{(x)})$ is the solution of

$$A_J u = -1, \quad u(j_1) = u(j_2) = 0.$$

From $A_J u = -1$, i.e., $u'' = -2$, it follows that $u = -x^2 + cx + d$, and from the boundary condition $u(j_1) = u(j_2) = 0$ we can determine c and d to obtain

$$u = (j_2 - x)(x - j_1).$$

The canonical measure dm equals

$$dm = 2dx \quad \text{since} \quad m = -\frac{dp_J}{ds} = -\frac{dp_J}{dx} = 2x - (j_1 + j_2).$$

Also, to obtain $\hat{\varphi}_\lambda(x) = E(e^{-\lambda\tau_J^{(x)}})$ introduced in the preceding section, we solve

$$\frac{1}{2}\lambda u'' = \lambda u, \quad u(j_1) = u(j_2) = 1.$$

This gives us

$$u = \alpha e^{\sqrt{2\lambda}x} + \beta e^{-\sqrt{2\lambda}x},$$

and the constants α and β can be determined by using the boundary condition $u(j_1) = u(j_2) = 1$.

5.11. Classification of Boundary Points with Respect to Feller's Operator $D_m D_s^+$

In this and the following few sections, we will leave Markov processes aside, and we only assume that on an interval $I = (r_1, r_2)$ a continuous, strictly increasing function $s(x)$ and a right continuous, increasing function $m(x)$ are given. We define the operator $(D_m D_s^+)_I$ using these functions as we have already done, and call it the **Feller's Operator**. Let us classify the end points r_1, r_2 of I with respect to $D_m D_s^+$ into the following four classes: **regular boundary, exit boundary, entrance**

boundary, and **natural boundary**. For this purpose, let us first introduce the following quantities:

$$\sigma_1 = \iint_{r_1<y<x<r_1'} dm(x)ds(y), \quad \mu_1 = \iint_{r_1<y<x<r_1'} ds(x)dm(y),$$

$$\sigma_2 = \iint_{r_2>y>x>r_2'} dm(x)ds(y), \quad \mu_2 = \iint_{r_2>y>x>r_2'} ds(x)dm(y).$$

Using these quantities, we will call

$$\begin{aligned}
&r_i \text{ a regular boundary if } \sigma_i < \infty,\ \mu_i < \infty,\\
&r_i \text{ an exit boundary if } \sigma_i < \infty,\ \mu_i = \infty,\\
&r_i \text{ an entrance boundary if } \sigma_i = \infty,\ \mu_i < \infty,\\
&r_i \text{ a natural boundary if } \sigma_i = \infty,\ \mu_i = \infty.
\end{aligned}$$

Actually, the values of σ_i and μ_i depend on the choice of r_i'. However, whether they take finite values or ∞ is independent of the choice of r_i', and hence, the classification above is also independent of the choice of r_i'.

We will postpone until later the explanation of the probabilistic meaning of this classification, and let us first give some examples of the four kinds of boundaries.

EXAMPLE 5.11.1. Let $D_m D_s^+ = \frac{d^2}{dx^2}$, $I = (-\infty, \infty)$. Then we have $m = x$ and $s = x$ so that

$$\sigma_2 = \iint_{\infty>y>x>r_2'} dxdy = \int_{r_2'}^{\infty} (y - r_2')dy = \infty,$$

$$\mu_2 = \iint_{\infty>y>x>r_2'} dxdy = \infty.$$

Therefore, ∞ is a natural boundary. Similarly, we can show that $-\infty$ is also a natural boundary.

EXAMPLE 5.11.2. If $D_m D_s^+ = \frac{d^2}{dx^2}$, $I = (-\infty, 0)$, then we can show that $-\infty$ is a natural boundary, while 0 is a regular boundary.

EXAMPLE 5.11.3. If $D_m D_s^+ = \frac{d}{dx} + x^2\frac{d^2}{dx^2}$, $I = (0, 2)$, then since

$$\frac{d}{dx} + x^2\frac{d^2}{dx^2} = \frac{d}{x^{-2}e^{-\frac{1}{x}}dx}\frac{d}{e^{\frac{1}{x}}dx},$$

we have

$$dm = x^{-2}e^{-\frac{1}{x}}dx, \quad ds = e^{\frac{1}{x}}dx.$$

Therefore, at 0, we have

$$\sigma = \iint_{0<y<x<1} x^{-2}e^{-\frac{1}{x}}e^{\frac{1}{y}}dxdy = \infty,$$

$$\mu = \iint_{0<y<x<1} y^{-2}e^{-\frac{1}{y}}e^{\frac{1}{x}}dydx < \infty,$$

which implies that 0 is an entrance boundary. Similarly, we can show that 2 is a regular boundary.

EXAMPLE 5.11.4. If $D_m D_s^+ = -\frac{d}{dx} + x^2\frac{d^2}{dx^2}$, $I = (0, \infty)$, then we can show that 0 is an exit boundary and ∞ is a natural boundary.

5.12. Particular Solutions of the Homogeneous Equation $(\lambda - D_m D_s^+)u = 0$ $(\lambda > 0)$

For convenience, let us suppose that $r_1 < 0 < r_2$ is satisfied. Furthermore, we suppose that

$$m(0-) = m(0+) = 0, \quad s(0) = 0.$$

A general solution of this homogeneous equation is a linear combination of the following particular solutions e_0 and e_1:

$$D_m D_s^+ e_0 = \lambda e_0, \quad e_0(0) = 1, \quad D_s^+ e_0(0) = 0, \tag{5.12.1}$$

$$D_m D_s^+ e_1 = \lambda e_1, \quad e_1(0) = 0, \quad D_s^+ e_1(0) = 1. \tag{5.12.2}$$

To find these particular solutions e_0 and e_1, it suffices to solve the following integral equations:

$$e_0(x) = 1 + \lambda \int_0^x \int_0^\xi e_0(\eta) dm(\eta) ds(\xi),$$
$$e_1(x) = s(x) + \lambda \int_0^x \int_0^\xi e_1(\eta) dm(\eta) ds(\xi).$$

Since it would be similar for the case $x < 0$, we consider only for the case $x > 0$, and try to find $e_i(x)$ in this case. Suppose we set

$$p_1 \circ p_2 \circ \cdots \circ p_n(x) = \int_{0 \le \xi_1 \le \cdots \le \xi_n \le x} \cdots \int dp_1(\xi_1) \cdots dp_n(\xi_n).$$

Then clearly the relation

$$(p_1 \circ p_2 \circ \cdots \circ p_m) \circ (p_{m+1} \circ \cdots \circ p_{m+n}) = p_1 \circ p_2 \circ \cdots \circ p_{m+n}$$

holds. Note, however, that the relations

$$(p_1 \circ p_2) \circ p_3 = p_1 \circ (p_2 \circ p_3), \quad p_1 \circ p_2 = p_2 \circ p_1$$

do not hold. Using this notation, we can rewrite the integral equations above in the following form:

$$e_0 = 1 + \lambda e_0 \circ m \circ s, \tag{5.12.3}$$

$$e_1 = s + \lambda e_1 \circ m \circ s. \tag{5.12.4}$$

Using the successive approximation method, we obtain formal expansions of e_0 and e_1 in the following form:

$$e_0 = 1 + \lambda m \circ s + \lambda^2 m \circ s \circ m \circ s + \lambda^3 m \circ s \circ m \circ s \circ m \circ s + \cdots,$$
$$e_1 = s + \lambda s \circ m \circ s + \lambda^2 s \circ m \circ s \circ m \circ s + \lambda^3 s \circ m \circ s \circ m \circ s \circ m \circ s + \cdots.$$

If we can show these expansions actually converge, then we will have the solutions to the equations above.

Let us put

$$\sigma = m \circ s, \quad \mu = s \circ m.$$

Then the quantities σ_2 and μ_2 that were introduced in the preceding section may be taken to be $\sigma(r_2)$ and $\mu(r_2)$, respectively, by using the functions above. Therefore, we have

$$\begin{aligned}
&\sigma(r_2) < \infty,\ \mu(r_2) < \infty, && \text{if } r_2 \text{ is a regular boundary,}\\
&\sigma(r_2) < \infty,\ \mu(r_2) = \infty, && \text{if } r_2 \text{ is an exit boundary,}\\
&\sigma(r_2) = \infty,\ \mu(r_2) < \infty, && \text{if } r_2 \text{ is an entrance boundary,}\\
&\sigma(r_2) = \infty,\ \mu(r_2) = \infty, && \text{if } r_2 \text{ is a natural boundary.}
\end{aligned}$$

By using the mathematical induction, we can show that

$$0 \le \underbrace{m \circ s}_{(1)}\underbrace{m \circ s}_{(2)} \circ \cdots \circ \underbrace{m \circ s}_{(n)}(x) \le \frac{\sigma(x)^n}{n!},$$

$$0 \le s \circ \underbrace{m \circ s}_{(1)}\underbrace{m \circ s}_{(2)} \circ \cdots \circ \underbrace{m \circ s}_{(n)}(x) \le \frac{s(x)\sigma(x)^n}{n!}.$$

Therefore, the series expansions for e_0 and e_1 both converge, and we have

$$e_0(x) \le e^{\lambda\sigma(x)}, \tag{5.12.5}$$

$$e_1(x) \le s(x) \cdot e^{\lambda\sigma(x)}. \tag{5.12.6}$$

Also, it is easy to see that we have

$$e_0(x) \ge \lambda m \circ s(x) = \lambda\sigma(x), \tag{5.12.7}$$

$$e_1(x) \ge \lambda s \circ m \circ s(x). \tag{5.12.8}$$

Using these inequalities, let us investigate the limiting values of $e_i(x)$ at the boundary r_2.

First of all, we note the following:

$$\begin{aligned}
&r_2 \text{ is regular} \longrightarrow m(r_2) < \infty,\ s(r_2) < \infty,\\
&r_2 \text{ is exit} \longrightarrow m(r_2) = \infty,\ s(r_2) < \infty,\\
&r_2 \text{ is entrance} \longrightarrow m(r_2) < \infty,\ s(r_2) = \infty,\\
&r_2 \text{ is natural} \longrightarrow m(r_2) \le \infty,\ s(r_2) \le \infty,
\end{aligned}$$

where in the case of a natural boundary, at least one of $m(r_2)$ and $s(r_2)$ must be ∞.

In order to see this, we note that

$$\sigma(r_2) = \int_0^{r_2} m(y)ds(y),$$

from which it follows that $\sigma(r_2) \ge m(r_2') \cdot (s(r_2) - s(r_2'))$, if $0 < r_2' < r_2$. Hence, if $s(r_2) = \infty$, then $\sigma(r_2) = \infty$, which in turn implies that if r_2 is a regular or an exit boundary, then $s(r_2) < \infty$. Similarly, we can conclude that if r_2 is a regular or an entrance boundary, then $m(r_2) < \infty$.

Also, from

$$\mu(r_2) = \int_0^{r_2} s(y)dm(y),$$

it follows that $\mu(r_2) \le s(r_2) \cdot m(r_2)$. Since $s(r_2) < \infty$ and $\mu(r_2) = \infty$ hold if r_2 is an exit boundary, we must have $m(r_2) = \infty$ in this case. Similarly, we can conclude that $s(r_2) = \infty$ if r_2 is an entrance boundary. If both $m(r_2)$ and $s(r_2)$ are finite, then it is easy to see that both $\mu(r_2)$ and $\sigma(r_2)$ are also finite, so r_2 is a regular boundary, and hence we can also conclude that in case r_2 is a natural boundary, then at least one of $m(r_2)$ or $s(r_2)$ must be ∞. An example for which $m(r_2) = \infty$, $s(r_2) < \infty$, and yet r_2 is a natural boundary can be obtained by taking

$$r_2 = 1, \quad s(x) = x, \quad m(x) = (1-x)^{-1},$$

while an example for which $m(r_2) < \infty$, $s(r_2) = \infty$, and with r_2 being a natural boundary can be obtained by setting

$$r_2 = 1, \quad s(x) = (1-x)^{-1}, \quad m(x) = x.$$

Finally, the combination

$$r_2 = \infty, \quad s(x) = x, \quad m(x) = x$$

gives, as we have seen in the preceding section, the case where $m(r_2) = \infty$, $s(r_2) = \infty$ with r_2 being a natural boundary.

Now, if r_2 is a regular or an exit boundary, then from (5.12.5)–(5.12.6), we know that $\sigma(r_2), s(r_2) < \infty$, and therefore $e_0(r_2), e_1(r_2) < \infty$. Also, if r_2 is an entrance or a natural boundary, then we have $\sigma(r_2) = \infty$, and therefore, by (5.12.7) we conclude that

$$e_0(r_2) = \infty$$

holds. For $e_1(x)$ in these cases, we have by (5.12.8)

$$\begin{aligned} e_1(x) &\geq \lambda s \circ m \circ s(x) = \lambda \int_0^x \int_0^\xi s(\eta) dm(\eta) ds(\xi) \\ &\geq \lambda s(r_2') \int_{r_2'}^x \int_{r_2'}^\xi dm(\eta) ds(\xi) \\ &\to \lambda s(r_2') \cdot \sigma(r_2) = \infty. \end{aligned}$$

Next, in order to find $D_s^+ e_0(x)$ and $D_s^+ e_1(x)$, we formally differentiate term by term the series expansions for e_0 and e_1 to obtain

$$\begin{aligned} D_s^+ e_0(x) &= \lambda m + \lambda^2 m \circ s \circ m + \lambda^3 m \circ s \circ m \circ s \circ m + \cdots, \\ D_s^+ e_1(x) &= 1 + \lambda s \circ m + \lambda^2 s \circ m \circ s \circ m + \cdots. \end{aligned}$$

If we can show that these series converge uniformly over every compact subset of I, then we would have $D_s^+ e_0(x)$ and $D_s^+ e_1(x)$ equal to the sum of these convergent series, respectively. The proof of the desired convergence can be established in the same way as for the case of the series expansions for e_0 and e_1. Also, in the same way as for the case of $e_i(r_2)$, we can establish

$$\begin{aligned} &r_2 \text{ is regular or entrance} \quad \to \quad D_s^+ e_i(r_2) < \infty, \ i = 1, 2, \\ &r_2 \text{ is exit or natural} \quad \to \quad D_s^+ e_i(r_2) = \infty, \ i = 1, 2. \end{aligned}$$

5.13. General Solutions of the Homogeneous Equation $(\lambda - D_m D_s^+)u = 0$ $(\lambda > 0)$

Let u_1, u_2 be two arbitrary solutions of

(5.13.1) $$(\lambda - D_m D_s^+)u = 0 \quad (\lambda > 0).$$

The quantity

$$W(x) = W(u_1, u_2)(x) = D_s^+ u_1(x) u_2(x) - D_s^+ u_2(x) u_1(x),$$

defined for this pair u_1, u_2, is called the **Wronskian** for u_1 and u_2.

THEOREM 5.13.1. *The Wronskian $W(u_1, u_2)(x)$ is independent of x.*

LEMMA 5.13.1. *Let u and v be functions of bounded variation. Then*

$$d(u \cdot v) = v^* du + u_* dv = v_* du + u^* dv \text{ holds,}$$

where dw is the signed measure corresponding to w, and $w^*(x) = w(x+0), w_*(x) = w(x-0)$.

PROOF. The assertion of the lemma clearly follows from

$$\begin{aligned} &u(x_i)v(x_i) - u(x_{i-1})v(x_{i-1}) \\ &\quad = v(x_i)(u(x_i) - u(x_{i-1})) + u(x_{i-1})(v(x_i) - v(x_{i-1})) \\ &\quad = v(x_{i-1})(u(x_i) - u(x_{i-1})) + u(x_i)(v(x_i) - v(x_{i-1})). \end{aligned}$$ □

PROOF OF THE THEOREM. By using the lemma, we have

$$\begin{aligned} dW &= u_2 d(D_s^+ u_1) + D_s^+ u_1 \cdot du_2 - u_1 d(D_s^+ u_2) - D_s^+ u_2 \cdot du_1 \\ &= u_2 \lambda u_1 dm + D_s^+ u_1 D_s^+ u_2 ds - u_1 \lambda u_2 dm - D_s^+ u_2 D_s^+ u_1 ds \\ &= 0. \end{aligned}$$

In particular, for e_0 and e_1 of the preceding section, we have

$$W(e_1, e_0) = W(e_1, e_0)(0) = (D_s^+ e_1 \cdot e_0 - D_s^+ e_0 \cdot e_1)(0) = 1.$$ □

LEMMA 5.13.2. *$e_0/e_1 > D_s^+ e_0 / D_s^+ e_1$ holds for $x > 0$.*

PROOF.

$$\frac{e_0}{e_1} - \frac{D_s^+ e_0}{D_s^+ e_1} = \frac{W(e_1, e_0)}{e_1 D_s^+ e_1} = \frac{1}{e_1 D_s^+ e_1} > 0.$$ □

LEMMA 5.13.3. *As x increases, e_0/e_1 decreases.*

PROOF.

$$D_s^+ \left[\frac{e_0}{e_1} \right] = \frac{-W(e_1, e_0)}{e_1^2} = -\frac{1}{e_1^2} < 0.$$ □

LEMMA 5.13.4. *As x increases, $D_s^+ e_0 / D_s^+ e_1$ increases.*

PROOF.

$$\begin{aligned} D_m \left(\frac{D_s^+ e_0}{D_s^+ e_1} \right) &= \frac{D_m D_s^+ e_0 \cdot D_s^+ e_1 - D_m D_s^+ e_1 \cdot D_s^+ e_0}{(D_s^+ e_1)^2} \\ &= \frac{\lambda W(e_1, e_0)}{(D_s^+ e_1)^2} = \frac{\lambda}{(D_s^+ e_1)^2} > 0. \end{aligned}$$ □

LEMMA 5.13.5. *As $x \uparrow r_2$,*

$$\frac{e_0}{e_1} - \frac{D_s^+ e_0}{D_s^+ e_1} = \frac{1}{e_1 D_s^+ e_1} \begin{cases} \to 0, & \textit{if } r_2 \textit{ is not regular,} \\ \to c > 0, & \textit{if } r_2 \textit{ is regular.} \end{cases}$$

Keeping these preliminary results in mind, let us now try to find solutions of the homogeneous equation (5.13.1) which satisfy the following two conditions:

(A) $u > 0$,

(B) $u \downarrow$ when $x \uparrow$.

By (A), we must have $u(0) > 0$. Hence, it is enough to find a solution with $u(0) = 1$, and multiply that with positive constants. So, let us find a solution of the form

$$u = e_0 - \gamma e_1.$$

By condition (A), we need

$$\gamma < \frac{e_0}{e_1} \quad (x > 0).$$

By condition (B), we need $D_s^+ u < 0$ so that we must have

$$\gamma > \frac{D_s^+ e_0}{D_s^+ e_1} \quad (x > 0).$$

By the lemmas obtained above, we know that

$$\overline{\gamma} = \lim_{x \uparrow r_2} \frac{e_0}{e_1}, \quad \underline{\gamma} = \lim_{x \uparrow r_2} \frac{D_s^+ e_0}{D_s^+ e_1} \tag{5.13.2}$$

both exist and satisfy $\overline{\gamma} \geq \underline{\gamma}$, and hence we must require that our γ satisfies

$$\overline{\gamma} \geq \gamma \geq \underline{\gamma}. \tag{5.13.3}$$

Conversely, if γ satisfies condition (5.13.3), then if we set

$$u = e_0 - \gamma e_1,$$

then it satisfies conditions (A) and (B) if $x > 0$. If $x < 0$, then the expansions of e_0 and e_1 tell us that both e_0 and $-e_1$ are positive and decreasing in this case, so that $u = e_0 - \gamma e_1$ again gives us the desired solution. Summarizing the results obtained so far, we can now state:

THEOREM 5.13.2. *Decreasing and positive solutions of the homogeneous equation above with the additional condition $u(0) = 1$ are given by*

$$u = e_0 - \gamma e_1,$$

where γ is an arbitrary positive number lying between $\underline{\gamma}$ and $\overline{\gamma}$. If r_2 is a regular boundary, $\underline{\gamma} < \overline{\gamma}$ holds, and hence there are infinitely many solutions in this case, and they all lie between $\overline{u} = e_0 - \underline{\gamma} e_1$ and $\underline{u} = e_0 - \overline{\gamma} e_1$. In all other cases, we have $\underline{\gamma} = \overline{\gamma}$, and the solution u is unique.

Let us next investigate the value of the solution u obtained in the theorem above at the boundary r_2. If r_2 is a regular boundary, then

$$\overline{u}(r_2) = e_0(r_2) - \frac{D_s^+ e_0(r_2)}{D_s^+ e_1(r_2)} e_1(r_2) = \frac{1}{D_s^+ e_1(r_2)},$$

$$\underline{u}(r_2) = e_0(r_2) - \frac{e_0(r_2)}{e_1(r_2)} e_1(r_2) = 0.$$

If r_2 is an exit boundary, we also have $u(r_2) = 0$. In case r_2 is an entrance boundary, we have

$$\begin{aligned} u(r_2) &= \lim_{x \uparrow r_2} \left\{ e_0(x) - \frac{D_s^+ e_0}{D_s^+ e_1}(r_2) e_1(x) \right\} \\ &\leq \lim_{x \uparrow r_2} \left\{ e_0(x) - \frac{D_s^+ e_0(x)}{D_s^+ e_1(x)} e_1(x) \right\} = \frac{1}{D_s^+ e_1(r_2)}. \end{aligned}$$

On the other hand, for x sufficiently close to r_2, we have

$$u(r_2) + \epsilon > e_0(x) - \frac{D_s^+ e_0}{D_s^+ e_1}(r_2) e_1(x),$$

while if y is also sufficiently close to r_2, then

$$u(r_2) + 2\epsilon > e_0(x) - \frac{D_s^+ e_0(y)}{D_s^+ e_1(y)} e_1(x).$$

Now since

$$D_s^+ \left\{ e_0(x) - \frac{D_s^+ e_0(y)}{D_s^+ e_1(y)} e_1(x) \right\} = \left\{ \frac{D_s^+ e_0(x)}{D_s^+ e_1(x)} - \frac{D_s^+ e_0(y)}{D_s^+ e_1(y)} \right\} D_s^+ e_1(x) < 0$$

by Lemma 5.13.4, we have that if $r_2 > y > x$ and x is sufficiently close to r_2, then

$$e_0(x) - \frac{D_s^+ e_0(y)}{D_s^+ e_1(y)} e_1(x) > e_0(y) - \frac{D_s^+ e_0(y)}{D_s^+ e_1(y)} e_1(y) = \frac{1}{D_s^+ e_1(y)},$$

which, combined with the inequality above, yields that

$$u(r_2) + 2\epsilon > \frac{1}{D_s^+ e_1(y)}.$$

By letting $y \uparrow r_2$, and then letting $\epsilon \downarrow 0$, we obtain

$$u(r_2) \geq \frac{1}{D_s^+ e_1(r_2)}.$$

Combined with the reverse inequality we already obtained, we finally conclude that

$$u(r_2) = \frac{1}{D_s^+ e_1(r_2)}$$

if r_2 is an entrance boundary. If r_2 is a natural boundary, we get the same equality as above, but since $D_s^+ e_1(r_2) = \infty$ in this case, we have $u(r_2) = 0$.

Similar arguments yield the values of $D_s^+ u(r_2)$ for every type of boundary, and combining these with the ones obtained above, we obtain the following theorem.

THEOREM 5.13.3. *For the solution u obtained in Theorem 5.13.2 and for its derivative $D_s^+ u$, we have the following limiting values at the boundary r_2:*

$$u(r_2) = \begin{cases} 0, & \text{if } r_2 \text{ is exit, natural, or regular (but only for } \underline{u}), \\ \frac{1}{D_s^+ e_1(r_2)}, & \text{if } r_2 \text{ is entrance or regular (but only for } \overline{u}). \end{cases}$$

$$D_s^+ u(r_2) = \begin{cases} 0, & \text{if } r_2 \text{ is entrance or natural or regular (but only for } D_s^+ \overline{u}), \\ -\frac{1}{e_1(r_2)}, & \text{if } r_2 \text{ is exit or regular (but only for } D_s^+ \underline{u}). \end{cases}$$

Since the solutions u and e_0 are linearly independent, arbitrary solutions to the given equation are given by linear combinations $\alpha u + \beta e_0$. Since both $u(r_2)$ and $D_s^+ u(r_2)$ are finite, a solution $\alpha u + \beta e_0$ takes the value ∞ only when $e_0(r_2) = \infty$. Therefore, we have the the following.

THEOREM 5.13.4. *A solution v of the homogeneous equation under consideration, which is different from u obtained above, satisfies the following:*

$$\begin{aligned} &r_2 \text{ is regular or exit} \rightarrow |v(r_2)| < \infty, \\ &r_2 \text{ is entrance or natural} \rightarrow |v(r_2)| = \infty, \\ &r_2 \text{ is regular or entrance} \rightarrow |D_s^+ v(r_2)| < \infty, \\ &r_2 \text{ is exit or natural} \rightarrow |D_s^+ v(r_2)| = \infty. \end{aligned}$$

5.14. Solutions of the Non-Homogeneous Equation $(\lambda - D_m D_s^+)g = f$

The solution u of the homogeneous equation obtained in the preceding section was a positive, decreasing solution, and it was shown that both u and $D_s^+ u$ have finite limits at the boundary r_2. Let us denote this solution u by $u_2(x) = u_2(x;\lambda)$. Similarly, we can show that there exists a positive, increasing solution $u_1(x) = u_1(x;\lambda)$, and that both u_1 and $D_s^+ u_1$ have finite limits at the boundary r_1.

As we proved in the preceding section, the Wronskian $W = W(u_1, u_2)$ is a constant, and by considering $W(0)$, for instance, we see that $W > 0$. Therefore, by multiplying one of the functions, say u_1, by a positive constant if necessary, we may assume that $W = 1$.

Now, let us define

$$K(x,y) = K(x,y;\lambda) = \begin{cases} u_1(x)u_2(y), & \text{for} \quad r_1 < x \le y < r_2, \\ u_2(x)u_1(y), & \text{for} \quad r_1 < y \le x < r_2. \end{cases}$$

Then K is a continuous function of two variables x and y, and is symmetric with respect to x and y. It is also clear that $K \ge 0$.

In order to find a particular solution g_0 of the inhomogeneous equation

$$(\lambda - D_m D_s^+)g = f \quad (\lambda > 0), \tag{5.14.1}$$

we set

$$g_0(x) = Kf(x) = \int_{r_1}^{r_2} K(x,y)f(y)dm(y),$$

where f is assumed to be a bounded and continuous function. $g_0(x)$ is well defined (i.e., the defining integral makes sense) and it is continuous in x as can be seen from the following:

$$\begin{aligned} g_0(x) &= u_2(x)\int_{r_1}^{x+0} f(y)u_1(y)dm(y) + u_1(x)\int_{x+0}^{r_2} f(y)u_2(y)dm(y) \\ &= \frac{u_2(x)}{\lambda}\int_{r_1}^{x+0} f(y)du_1^+(y) + \frac{u_1(x)}{\lambda}\int_{x+0}^{r_2} f(y)du_2^+(y), \end{aligned}$$

where we wrote u_i^+ for $D_s^+ u_i$.

Thus, the integral operator K is well defined, and it is positive and linear.

In order to show that g_0 thus defined satisfies equation (5.14.1), we use Lemma 5.13.1 of the preceding section and obtain

$$\begin{aligned} dg_0 &= du_2 \int_{r_1}^{x+0} f(y)u_1(y)dm(y) + u_2(x)f(x)u_1(x)dm \\ &\quad + du_1 \int_{x+0}^{r_2} f(y)u_2(y)dm(y) - u_1(x)f(x)u_2(x)dm \\ &= du_2 \int_{r_1}^{x+0} f(y)u_1(y)dm(y) + du_1 \int_{x+0}^{r_2} f(y)u_2(y)dm(y), \end{aligned}$$

$$D_s^+ g_0 = D_s^+ u_2 \int_{r_1}^{x+0} f(y)u_1(y)dm(y) + D_s^+ u_1 \int_{x+0}^{r_2} f(y)u_2(y)dm(y).$$

Repeating the same kind of calculations and using the identities $\lambda u_i = D_m D_s^+ u_i$, $i = 1, 2$, and the fact $D_s^+ u_1 \cdot u_2 - D_s^+ u_2 \cdot u_1 = W(u_1, u_2) = 1$, we finally obtain

$$D_m D_s^+ g_0 = \lambda g_0 - f.$$

If either one or both of r_1 and r_2 are regular boundaries, then there are infinitely many pairs of corresponding solutions u_1, u_2, and accordingly, K and g_0 can be defined in infinitely many different ways. Nevertheless, every K and g_0 thus defined retain the properties described above.

If we investigate the limiting values of g_0 at the boundary points, we obtain the following theorem. For the sake of simplicity, we write $g_0(r_2)$ for $g_0(r_2-)$.

THEOREM 5.14.1. (A) *If r_2 is a regular boundary, then*

$$g_0(r_2) = u_2(r_2)\int_{r_1}^{r_2} fu_1 dm,$$

$$g_0^+(r_2) = u_2^+(r_2)\int_{r_1}^{r_2} fu_1 dm.$$

(B) *If r_2 is an exit boundary, then*

$$g_0(r_2) = 0.$$

(C) *If r_2 is an entrance boundary, then*

$$g_0(r_2) = u_2(r_2)\int_{r_1}^{r_2} fu_1 dm,$$

$$g_0^+(r_2) = 0.$$

(D) *If r_2 is a natural boundary, then $g_0(r_2)$ exists as long as $f(r_2)$ exists, and*

$$g_0(r_2) = \frac{f(r_2)}{\lambda}.$$

PROOF. Assertion (A) is obvious.

For the proof of (B), we note that since

$$|g_0(x)| \leq (K\cdot 1)(x)\cdot \sup_x |f(x)|$$

is satisfied, it is sufficient to show that $(K\cdot 1)(x) \to 0$ as $x \to r_2$. This can be established as follows:

$$\begin{aligned}
\lambda(K\cdot 1)(x) &= u_2(x)\int_{r_1}^{x+0} du_1^+ + u_1(x)\int_{x+0}^{r_2} du_2^+ \\
&= u_2(x)u_1^+(x) - u_2(x)u_1^+(r_1) + u_1(x)[u_2^+(r_2) - u_2^+(x)] \\
&= u_2(x)u_1^+(x) + o(1) \quad \text{(as } r_2 \text{ is an exit boundary)} \\
&= u_1^+(x)\int_x^{r_2} (-u_2^+)ds + o(1) \\
&\leq -u_2^+(x)\int_x^{r_2} u_1^+(y)ds + o(1) \quad (\text{since } -u_2^+\downarrow,\ u_1^+\uparrow) \\
&= -u_2^+(x)(u_1(r_2) - u_1(x)) + o(1) \to 0.
\end{aligned}$$

For the proof of (C), we see that by letting $M = \sup_{r_1\leq y\leq r_2}|f(y)|$ and since $u_2^+(r_2) = 0$, we have

$$\left| u_1(x)\int_{x+0}^{r_2} f(y)du_2^+(y)\right| \leq -Mu_1(x)u_2^+(x) = M\left(1 - u_1^+(x)u_2(x)\right).$$

Observe that u_1 admits the following expression for some $\tilde{\gamma} > 0$:

$$u_1(x) = \frac{1}{\gamma + \tilde{\gamma}}(e_0(x) + \tilde{\gamma} e_1(x)).$$

Therefore,

$$u_2(r_2)u_1^+(r_2) = \frac{1}{\gamma + \tilde{\gamma}}\left\{\frac{e_0^+(r_2)}{e_1^+(r_2)} + \tilde{\gamma}\right\} = 1,$$

and

$$\lim_{x \to r_2} u_1(x) \int_{x+0}^{r_2} f(y) du_2^+(y) = 0.$$

The second half of assertion (C) is clear.

For the proof of (D), notice that $u_2(r_2) = 0$ and $u_2^+(r_2) = 0$. For any $\epsilon > 0$, there exists ξ such that, for any $x > \xi$, $|f(x) - f(r_2)| < \epsilon$. We have then for any $x > \xi$

$$\begin{aligned} g_0(x) &= u_2(x) \int_{r_1}^{\xi} f(y)u_1(y)dm(y) \\ &\quad + \frac{u_2(x)}{\lambda} \int_{\xi}^{x+0} f(y)du_1^+(y) + \frac{u_1(x)}{\lambda} \int_{x+0}^{r_2} f(y)du_2^+(y) \\ &\leq u_2(x) \int_{r_1}^{\xi} f(y)u_1(y)dm(y) \\ &\quad + \frac{f(r_2)+\epsilon}{\lambda}\left(u_2(x)u_1^+(x) - u_2(x)u_1^+(\xi) - u_1(x)u_2^+(x)\right) \\ &= u_2(x) \int_{r_1}^{\xi} f(y)u_1(y)dm(y) + \frac{f(r_2)+\epsilon}{\lambda}\left(1 - u_2(x)u_1^+(\xi)\right). \end{aligned}$$

Letting $x \to r_2$, we get $g_0(r_2) \leq \frac{f(r_2)+\epsilon}{\lambda}$. Similarly, we can get $g_0(r_2) \geq \frac{f(r_2)-\epsilon}{\lambda}$. □

EXAMPLE 5.14.1. Let $I = (-\infty, \infty)$ and $D_m D_s^+ = \frac{1}{2}\frac{d^2}{dx^2}$. Then both $-\infty$ and ∞ are natural boundaries. Hence u_1, u_2 can be determined uniquely up to a multiplicative constant. Indeed, if we solve

$$\frac{1}{2}\frac{d^2}{dx^2}u_i = \lambda u_i, \quad u_i > 0, \quad u_1 \uparrow, \quad u_2 \downarrow,$$

then we obtain

$$u_1 = e^{\sqrt{2\lambda}x}, \quad u_2 = e^{-\sqrt{2\lambda}x},$$

for which $W(u_1, u_2) = 2\sqrt{2\lambda}$. So, if we take $u_1/(2\sqrt{2\lambda})$ instead of u_1, we have a pair of solutions for which $W = 1$. Then we can write

$$K(x, y) = \frac{1}{2\sqrt{2\lambda}} e^{-\sqrt{2\lambda}|x-y|},$$

and since $s = x$ and $m = 2x$, we get

$$Kf = \frac{1}{\sqrt{2\lambda}} \int_{-\infty}^{\infty} e^{-\sqrt{2\lambda}|x-y|} f(y)dy.$$

5.15. Distributions of Various Quantities Associated with $x^{(a)}(t)$ in a Regular Interval

In the previous few sections, we investigated only analytic properties of the Feller operator $D_m D_s^+$. Let us now go back to the investigation of our original problem. Let $x^{(a)}(t)$ be a Markov process moving on a compact metric space R, and suppose an open subset I of R is a regular interval for $x^{(a)}(t)$. This means, of course, that I is homeomorphic to an interval (r_1, r_2) of the reals and that every point of I is not only a diffusive point but also a regular point. We have explained that there exist the canonical scale s and the canonical measure dm defined on I and that we can represent

$$A_I = (D_m D_s^+)_I. \tag{5.15.1}$$

As before, we identify points of I with the corresponding points of (r_1, r_2). Let us denote by $P_I(t, \xi, E)$ the probability for a process starting from the point ξ of I to reach the set E after time t without leaving the interval I. If we denote by $\tau_I^{(\xi)}$ the first instant when $x^{(\xi)}(t)$ leaves I, then we can write

$$P_I(t, \xi, E) = P\{x^{(\xi)}(t) \in E,\ \tau_I^{(\xi)} > t\}. \tag{5.15.2}$$

Furthermore, if $\tau_I^{(\xi)} < \infty$, then $x^{(\xi)}(\tau_I^{(\xi)})$ does not belong to I itself, but is a point of the boundary ∂I of I in R. However, for an arbitrary $\delta > 0$, the point $x^{(\xi)}(\tau_I^{(\xi)} - \delta)$ lies in I, and since the sample process of $x^{(\xi)}(t)$ has, at most, points of discontinuity of the first kind, $x^{(\xi)}(\tau_I^{(\xi)} - 0)$ can be determined and must coincide either with r_1 or r_2. We set

$$\tau_1^{(\xi)} = \tau_1^{(\xi)}(I) = \begin{cases} \tau_I^{(\xi)}, & \text{if } \tau_I^{(\xi)} < \infty \text{ and } x^{(\xi)}(\tau_I^{(\xi)} - 0) = r_1, \\ \infty, & \text{otherwise.} \end{cases}$$

Similarly, by using r_2 in place of r_1, we define $\tau_2^{(\xi)} = \tau_2^{(\xi)}(I)$. Then it is clear that $\tau_i^{(\xi)}$ represents the first time when $x^{(\xi)}(t)$ leaves I through the end point r_i. Denote by $\varphi_i(dt, \xi)$ the distribution of $\tau_i^{(\xi)}$. Instead of computing $P_I(t, \xi, E)$ and $\varphi_i(dt, \xi)$ directly, let us determine their Laplace transforms:

$$R_I(\lambda, x, E) = \int_0^\infty e^{-\lambda t} P_I(t, x, E)\,dt \tag{5.15.3}$$

and

$$\hat{\varphi}_{i\lambda}(\xi) = \int_0^\infty e^{-\lambda t} \varphi_i(dt, \xi). \tag{5.15.4}$$

The main objective of this section is to prove the following theorem.

THEOREM 5.15.1. *We have*

$$R_I(\lambda, x, E) = \int_E K_\lambda(x, y)\,m(dy), \tag{5.15.5}$$

$$\hat{\varphi}_{1\lambda}(\xi) = \frac{u_{2\lambda}(\xi)}{u_{2\lambda}(r_1)}, \quad \hat{\varphi}_{2\lambda}(\xi) = \frac{u_{1\lambda}(\xi)}{u_{1\lambda}(r_2)}, \tag{5.15.6}$$

where $u_{i\lambda}(\xi)$, $i = 1, 2$, *and* $K_\lambda(x, y) = K(x, y, \lambda)$ *are the functions introduced in the preceding section.*

REMARK. In case r_i is a regular boundary, we will take, in the theorem above, the solution $\underline{u}_{i\lambda}(\xi)$, i.e., the solution satisfying the condition $u_{i\lambda}(r_i) = 0$, and $K_\lambda(x, y)$ to be the one constructed from this particular $u_{i\lambda}$.

Before proving the theorem, we observe the following. If $u_{2\lambda}(r_1) = \infty$, then from (5.15.6) it follows that $\hat{\varphi}_{1\lambda}(\xi) = 0$, and therefore, $P(\tau_1^{(\xi)} = \infty) = 1$, which implies that the process $x^{(\xi)}(t)$ never reaches the point r_1. On the other hand, if $u_{2\lambda}(r_1) < \infty$, then $\hat{\varphi}_{1\lambda}(\xi) > 0$, and hence $P(\tau_1^{(\xi)} < \infty) > 0$, which means that the process $x^{(\xi)}(t)$ has a positive probability of reaching r_1. Thus we have the following theorem, which is a corollary to Theorem 5.15.1.

THEOREM 5.15.2. *It is possible for $x^{(\xi)}(t)$ to reach a regular boundary or an exit boundary within a finite length of time, but it is impossible to reach an entrance or a natural boundary within a finite length of time.*

PROOF OF THEOREM 5.15.1. First of all, we take a sub-interval $J = (j_1, j_2)$ with $\overline{J} \subset I$, and prove the theorem for J. We use the notation $R_J(\lambda, x, E)$, $K_\lambda(x, y; J)$, $\varphi_i(dt, \xi, J)$, $\hat{\varphi}_{i\lambda}(\xi; J)$, and $u_{i\lambda}(\xi; J)$ for the quantities defined for J. As we have shown in §4.12, the application of the strong Markov property gives us

$$R_\lambda f(x) = R_{J\lambda} f(x) + \sum_{i=1}^{2} \hat{\varphi}_{i\lambda}(x, J) R_\lambda f(j_i) \tag{5.15.7}$$

for $f \in \boldsymbol{C}(R)$. Since $\hat{\varphi}_{1\lambda}(x; J)$ is the solution of the boundary value problem

$$(\lambda - A_J)u = 0, \quad u(j_1) = 1, \quad u(j_2) = 0,$$

and since $A_J = (D_m D_s^+)_J$, we get

$$\hat{\varphi}_{1\lambda}(x; J) = \frac{u_{2\lambda}(x; J)}{u_{2\lambda}(j_1; J)}. \tag{5.15.8}$$

Similarly, we obtain

$$\hat{\varphi}_{2\lambda}(x; J) = \frac{u_{1\lambda}(x; J)}{u_{1\lambda}(j_2; J)}. \tag{5.15.9}$$

Therefore, we can write

$$R_\lambda f(x) = R_{J\lambda} f(x) + u_{1\lambda}(x; J) \frac{R_\lambda f(j_2)}{u_{1\lambda}(j_2; J)} + u_{2\lambda} \frac{R_\lambda f(j_1)}{u_{2\lambda}(j_1; J)}. \tag{5.15.10}$$

On the other hand, since $R_\lambda f$ satisfies the equation $(\lambda - A) R_\lambda f = f$, we have, naturally, $(\lambda - A_J) R_\lambda f = f$ in J, and since $A_J = (D_m D_s^+)_J$, we can use the result of the preceding section to obtain

$$R_\lambda f(x) = K_{\lambda J} f(x) + c_1 u_{1\lambda}(x; J) + c_2 u_{2\lambda}(x; J), \quad x \in J.$$

Since the end points of J are regular boundaries of J, we have

$$K_{\lambda J} f(j_i) = 0, \quad u_{i\lambda}(j_i; J) = 0, \quad i = 1, 2.$$

Therefore, we must have

$$c_1 = \frac{R_\lambda f(j_2)}{u_{1\lambda}(j_2; J)}, \quad c_2 = \frac{R_\lambda f(j_1)}{u_{2\lambda}(j_1; J)},$$

which enables us to write

$$R_\lambda f(x) = K_{\lambda J} f(x) + \frac{R_\lambda f(j_2)}{u_{1\lambda}(j_2; J)} u_{1\lambda}(x; J) + \frac{R_\lambda f(j_1)}{u_{2\lambda}(j_1; J)} u_{2\lambda}(x; J). \tag{5.15.11}$$

Comparison of (5.15.10) and (5.15.11) yields that

$$R_{J\lambda}f(x) = K_{\lambda J}f(x), \quad x \in J, \tag{5.15.12}$$

for $f \in \boldsymbol{C}(R)$. Since an arbitrary continuous function on $\overline{J}$ can be extended to a continuous function on all of R, we can conclude that the relation (5.15.12) must hold for all $f \in \boldsymbol{C}(J)$. Thus, we obtain

$$\begin{aligned} R_J(\lambda; x, dy) &= K_J(\lambda, x, y)dm(y) \\ &= \begin{cases} u_{1\lambda}(x; J)u_{2\lambda}(y; J)dm(y), & j_1 \le x \le y \le j_2, \\ u_{2\lambda}(x; J)u_{1\lambda}(y; J)dm(y), & j_1 \le y \le x \le j_2. \end{cases} \end{aligned}$$

Now, if we let $J \uparrow I$, then $\tau_J^{(\xi)} \uparrow \tau_I^{(\xi)}$, and since

$$R_{J\lambda}f(\xi) = E\left\{\int_0^{\tau_J^{(\xi)}} e^{-\lambda t} f(x(t))dt\right\},$$

we have $R_{J\lambda}f(\xi) \to R_{I\lambda}f(\xi)$. Although $u_{1\lambda}(x; J)$ and $u_{2\lambda}(x; J)$ are determined uniquely only up to constant factors, we can choose suitable constant factors to make $u_{1\lambda}(x; J)$ and $u_{2\lambda}(x; J)$ approach $u_{1\lambda}(x)$ and $u_{2\lambda}(x)$, respectively, as $J \uparrow I$. Therefore, we get $R_{I\lambda} = K_{\lambda I}$, which yields the identity (5.15.5). As for $\hat{\varphi}_{i\lambda}(\xi)$, we achieve $J \uparrow I$ by first letting $j_2 \uparrow r_2$ and then letting $j_1 \downarrow r_1$. We then see that from (5.15.8)–(5.15.9) the identity (5.15.6) follows easily. □

5.16. Behavior of a Process at the Boundaries of a Regular Interval

Let us follow the notation of the preceding section. Since I is homeomorphic to (r_1, r_2), I contains points which correspond to points arbitrarily close to r_1, but there is no point in I corresponding to r_1 itself. If $\tau_I^{(\xi)}$ is finite, then $x^{(\xi)}(\tau_I^{(\xi)})$ is a point on the boundary ∂I of I in R. Let b be a point in ∂I of I. Then there exists a sequence $\{b_n\}$ of points of I converging to b. The sequence of real numbers corresponding to $\{b_n\}$ (i.e., the sequence of coordinates of $\{b_n\}$) has either one or both of r_1 and r_2 as a limit point. Suppose now the sequence $\{b_n\}$ has r_i as a limit point. Then we say that "b corresponds to r_i". It is possible that a single point b may correspond to both r_1 and r_2, or that two different points b and b' correspond to the same r_i. The latter can occur only when r_i is a natural boundary (see Theorem 5.16.1 below). Suppose now both b and b' correspond to r_1. Then we have $b_n \to b$, $b'_n \to b'$ and $\bar{b}_n (=$ the coordinate of $b_n) \to r_1$, $\overline{b}'_n \to r_1$. As we saw in the preceding section, we have for some non-negative constants c_1, c_2,

$$R_\lambda f = K_\lambda f + c_1 u_1 + c_2 u_2 \quad \text{in } I.$$

If r_1 is an entrance boundary, we have $u_2(r_1 + 0) = \infty$; hence, we must have $c_2 = 0$ in order for the identity above to be valid. Since $K_\lambda f(r_1 + 0)$ and $u_1(r_1 + 0)$ exist, we have

$$R_\lambda f(b) = \lim_n R_\lambda f(b_n) = \lim_n \{K_\lambda f(b_n) + c_1 u_1(b_n)\} = K_\lambda f(r_1 + 0) + c_1 u_1(r_1 + 0).$$

Similarly, we have $R_\lambda f(b') = K_\lambda f(r_1 + 0) + c_1 u_1(r_1 + 0)$, and hence $R_\lambda f(b') = R_\lambda f(b)$. Using the fact that $\lambda R_\lambda f \to f$ as $\lambda \to \infty$, we then obtain $f(b') = f(b)$. Since this holds for an arbitrary $f \in \boldsymbol{C}(R)$, we then conclude that $b' = b$. If r_1 is either an exit or a regular boundary, then $u_2(r_1 + 0)$ is finitely determined, and hence the same argument as above yields that $b' = b$. Therefore, we have

THEOREM 5.16.1. *If r_i is not a natural boundary, then there exists only one point in the boundary of I which corresponds to r_i.*

(i) Let us investigate the behavior of $x^{(a)}(t)$ at the boundary points corresponding to a natural boundary. If r_1 is a natural boundary, then in general there exist many points of R which correspond to r_1, and the set of all such points forms a closed subset F of R. We can show, furthermore, that a process $x^{(a)}(t)$ starting from a point of F is always confined within F. In order to show this, let us take a continuous function f on R which is identically equal to 0 on some neighborhood of F. Then f vanishes in some neighborhood in I of the point r_1. Therefore, $f(r_1+0)$ exists and is equal to 0, which in turn shows that $R_\lambda f(r_1+0) = f(r_1+0)/\lambda = 0$. Consequently, we have $R_\lambda f(a) = 0$ for $a \in F$. Now, let f be a continuous function which is 0 on F, 1 outside of some neighborhood U of F, and takes values between 0 and 1 in $U - F$. Then, for this f and for $a \in F$, we have

$$R_\lambda f(a) \geq \int_0^\infty e^{-\lambda t} P(t, a, U^c) dt.$$

Since the left-hand side of the inequality above is 0, we must have $P(t, a, U^c) = 0$, i.e., $P(t, a, U) = 1$. By letting $U \downarrow F$, we obtain $P(t, a, F) = 1$, and since $x^{(a)}(t)$ is right continuous, we conclude that

$$P(x^{(a)}(t) \in F, 0 \leq t < \infty) = 1, \quad a \in F.$$

If F consists of a single point, in particular, then that point is a trap. In general, it is not possible for a process starting from a point of I to reach F within a finite length of time.

(ii) Next, consider the case of an exit boundary. Let r_1 be an exit boundary. Since there corresponds only one boundary point to r_1, let us denote this point also by r_1. Take a neighborhood U of r_1 and consider its boundary. Then it consists of a point r_1' of I and a closed set C disjoint from I. Let $\tau^{(a)}$ be the instant when a process starting from a point a in U reaches r_1' before reaching C. We set $\tau^{(a)} = \infty$ if this does not occur. If we define

$$U(a) = E\{e^{-\lambda \tau^{(a)}}\},$$

then this satisfies

$$\begin{aligned} (\lambda - D_m D_s^+) U(a) &= 0, \quad \text{if } a \in (r_1, r_1'), \\ U(r_1') &= 1, \\ U(x) \leq U(y), &\quad \text{if } r_1 \leq x < y < r_1'. \end{aligned}$$

Consequently, we have

$$U(r_1) \leq U(x) = \frac{u_{1\lambda}(x)}{u_{1\lambda}(r_1')} \quad (r_1 < x < r_1').$$

If r_1 is an exit boundary, then $u_{1\lambda}(x) \to 0$ as $x \to r_1$. Therefore, we have $U(r_1) = 0$. This means that a process leaving from r_1 cannot go inside of I from the side of r_1. As we have seen before, the probability for a process starting from points of I to reach r_1, however, is positive. This is the reason why r_1 is called an exit boundary.

(iii) Let us now consider the case of an entrance boundary. We already explained that it is impossible for a process starting from I to reach an entrance boundary. Since there is only one point in the boundary of I in R which corresponds to an entrance boundary r_1, let us denote this boundary point again by r_1.

If we suppose that r_1 is also a diffusive point for $x^{(a)}(t)$, then a process starting from r_1 enters I immediately, and then it can never come back to r_1 from I. This is the reason for the name entrance boundary.

Let us first show that r_1 cannot be a trap. For this purpose, take a function f, which is continuous on R, and $f \geq 0$ and takes values $f(r_1) = 0$ and $f(r_2') = 1$, where r_2' is an arbitrarily chosen fixed point in I. Then, for $r_1 < a < r_2'$, we have

$$R_\lambda f(a) \geq \hat{\varphi}_{2\lambda}(a) R_\lambda f(r_2'),$$

where $\hat{\varphi}_{2\lambda}(a)$ is defined with respect to the interval (r_1, r_2'). Therefore, $\hat{\varphi}_{2\lambda}(a) = u_{1\lambda}(r_2')/u_{1\lambda}(a)$ and this approaches the positive value $u_{1\lambda}(r_2')/u_{1\lambda}(r_1)$ as $a \to r_1$. Therefore, we have

$$R_\lambda f(r_1) \geq \frac{u_{1\lambda}(r_2')}{u_{1\lambda}(r_1)} R_\lambda f(r_2').$$

If r_1 is a trap, then the left-hand side of the inequality above is always equal to $f(r_1)/\lambda = 0$, while the right-hand side is positive for sufficiently large λ, since $\lambda R_\lambda f(r_2') \to f(r_2') = 1$. But this is a contradiction and shows that r_1 cannot be a trap. Furthermore, we can show that a process starting from r_1 enters I immediately with probability 1, but the proof of this fact is rather complicated, and so we omit it here. From these facts, we see that r_1 is a generalized right shunt.

Let us consider the local generator at r_1. Take a sufficiently small neighborhood U of r_1 so that $p_U(\xi) = E(\tau_U^{(\xi)}) < \infty$. Let us denote by $\tau(\xi)$ the first time the process starting at r_1 reaches ξ. Since r_1 is a shunt, as we remarked above, it is necessary for the process to go through the point ξ in order to go outside of U, and therefore, we have $\tau(\xi) < \tau_U^{(r_1)}$. If we write $p(\xi) = E(\tau(\xi))$, then we have

$$p_U(r_1) = p(\xi) + p_U(\xi).$$

But Lemma 5.7.2 gives us

$$A_J p_U(\xi) = -1 \quad (J = I \cap U),$$

so that we have $A_J p(\xi) = 1$, i.e., $D_m D_s^+ p(\xi) = 1$. From this it follows that

$$p(\xi) = \int_{\xi_0}^{\xi} m(x) ds(x) + as(\xi) + b,$$

where ξ_0 is a constant less than ξ and arbitrarily chosen. If we take ξ_0 to be r_1, however, the integral above diverges and the identity would not make sense. So, we have to avoid this choice. Actually, the expression above is not really convenient, because the constants a and b appearing in it depend on the choice of ξ_0, and because it is valid only when $\xi > \xi_0$. However, we can rewrite the expression above in the form

$$p(\xi) = \int_{r_1}^{\xi} (s(\xi) - s(x)) dm(x) + cs(\xi) + d,$$

which does not involve ξ_0 and is valid for all ξ. Since $p(r_1) = 0$, we must have $c = d = 0$, taking into account the fact that $s(r_1) = -\infty$. Thus, we can now write

$$p(\xi) = \int_{r_1}^{\xi} (s(\xi) - s(x)) dm(x),$$

and by virtue of Theorem 5.8.1, we finally obtain

$$A_{r_1} f = \lim_{\xi \downarrow r_1} \frac{f(\xi) - f(r_1)}{\int_{r_1}^{\xi} (s(\xi) - s(x)) dm(x)}.$$

(iv) There are several different possibilities for the behavior of $x^{(a)}(t)$ at a regular boundary r_1, and this diversity of the behavior reflects the diversity for the choice of the values of $u_{1\lambda}(x)$ at such a boundary point. Here, we will be concerned only with the case where r_1 is also on the boundary of R itself, and neighborhoods of r_1 consist only of r_1 and points from I. In such a case, r_1 becomes either a trap or a generalized right shunt. If it is a trap, then we have $R_\lambda f(x) = f(r_1)/\lambda$ so that we can write

$$R_\lambda f(x) = K_\lambda f(x) + \frac{u_{2\lambda}(x)}{u_{2\lambda}(r_1)} \cdot \frac{f(r_1)}{\lambda} + \frac{u_{1\lambda}(x)}{u_{1\lambda}(r_2)} \cdot R_\lambda f(r_2).$$

If r_1 is a generalized shunt, then we can compute $p(\xi)$ as in (iii) to get

$$p(\xi) = \int_{r_1}^{\xi} (m(x) + c) ds(x).$$

As additive constants are immaterial for $m(x)$, we may set $c = 0$ to get

$$p(\xi) = \int_{r_1}^{\xi} m(x) ds(x).$$

Since $p(\xi) \uparrow$, we must have $m(x) > 0$, which implies that

$$m(r_1) \geq 0.$$

We have also

$$A_{r_1} f = \lim_{\xi \downarrow r_1} \frac{f(\xi) - f(r_1)}{\int_{r_1}^{\xi} m(x) ds(x)}.$$

If f belongs not only to $\mathfrak{D}(A_{r_1})$, but also to $\mathfrak{D}(A_U)$ for some neighborhood U of r_1, then $f \in \mathfrak{D}((D_m D_s^+)_J)$, where $J = U \cap I$ so that we have

$$A_U f(r_1) = \lim_{\xi \downarrow r_1} \frac{D_s^+ f(\xi)}{m(\xi)}.$$

If $m(r_1) = 0$, then we must have

$$D_s^+ f(r_1) = 0.$$

This condition is called the **boundary condition for a reflecting barrier**. If, on the other hand, $m(r_1) > 0$, then the continuity of $A_U f(x)$ requires that we must have

$$\frac{D_s^+ f(r_1)}{m(r_1)} = D_m D_s^+ f(r_1).$$

This is called the **boundary condition for a generalized reflecting barrier** and it was first introduced by Feller.

Many interesting facts can be observed by investigating the probabilistic meaning of the boundary conditions introduced above.

Postscript

Let us first list books which are good references for the topics covered in this book as a whole.

- A. Kolmogoroff: Grundbegriffe der Wahrscheinlichkeitsrechnung, Erg. der Math. (Berlin, 1933).
- W. Feller: An introduction to probability theory and its applications (1950).
- J. L. Doob: Stochastic processes (1952).
- P. Lévy [1]: Théorie de l'addition des variables aléatoires (Paris, 1937). [2]: Processus stochastiques et mouvement brownien (Paris, 1948).
- Y. Kawada: Probability Theory (in Japanese) (Kyouritsu Shuppan, 1948).
- K. Kunisawa: Modern Probability Theory (in Japanese) (Iwanami Zensho, 1951).
- K. Itô: Probability Theory (in Japanese) (Gendai Sugaku, Iwanami, 1952).
- G. Maruyama: Probability Theory (in Japanese) (Kyoritsu Gendai Sugaku Koza, 1957).

Let me explain briefly the contents of each chapter.

Chapter 1: Fundamental concepts of the probability theory are introduced in this chapter. The most convenient way to construct a mathematically rigorous theory of probability is to adopt the measure theoretic method of A. Kolmogorov. We followed this method in this book, but it is essential, in order to get familiar with applications to practical problems, to understand a more intuitive background for probability theory. For acquiring such an intuitive background, we recommend the book by W. Feller quoted above.

As basic material related to the measure theoretic method of Kolmogorov, we recommend, in addition to the book by Kolmogorov, the book by Y. Kawada mentioned above. However, neither book treats the detailed description of sample processes of stochastic processes. In order to investigate such topics, the notion of separability introduced by J. L. Doob is indispensable. The original paper of Doob, which took up this concept for the first time, however, was not written with sufficient rigor and was hard to go through. But the book by Doob mentioned above treats this topic rigorously and efficiently, and is quite readable.

Chapter 2: Discussions in this chapter were centered on the topics of continuous time parameter stochastic processes with independent increments (additive processes). Discussion of the case of discrete parameter (additive sequences) is limited to the situations, which have some bearing on the investigation of additive processes. For instance, detailed discussions of laws of large numbers, central limit theorems, laws of iterated logarithms, and so on were omitted. For these topics, we refer the readers to the book by K. Kunisawa mentioned above. Also the English translation by K. L. Chung of the following book originally written in Russian gives a well-organized treatment of these topics:

- Gnedenko-Kolmogoroff: Limit distributions for sums of independent random variables (Moscow-Leningrad, 1949).

Concerning additive processes, the aforementioned books by P. Lévy have a wealth of material, but are difficult to read through. Basic parts of the theory covered in his books are explained from the view point of Kolmogorov's approach in the books by Doob and K. Itô mentioned above, and are more accessible.

Chapter 3: Basic facts concerning stationary processes have been treated in this chapter, but results on interpolation and extrapolation by N. Wiener and Kolmogorov, and on parameter estimation by U. Grenander, were omitted. We hope these topics would be taken up in the book by T. Kawata entitled "The Applications of Stochastic Processes" in this series. Let us list a few references on these topics.

- Chapter 12 of the book by Doob mentioned above.
- N. Wiener: Extrapolation, interpolation and smoothing of stationary time series (1949).
- A. Kolmogoroff: Interpolation und Extrapolation in stationären Zufälligen Folgen, Bull. Acad. Sci. U.R.S.S. Ser. Math. 5 (1941).
- U. Grenander: Stochastic processes and statistical inference, Arkiv. för Mat. 1 (1950).

As for the generalized harmonic analysis that we touched on without proofs in this book, we refer the readers to:

- N. Wiener: Generalized harmonic analysis, Acta. Math. 55 (1930).

Chapters 4 and 5: In these two chapters, we restricted ourselves to the discussion of temporally homogeneous Markov processes, and called them simply Markov processes. The reason for this restriction is that the temporally homogeneous case is where a more or less complete theory has been established. But even here, not everything worth mentioning has been covered. Chapter 4 dealt only with basic material on Markov processes, and in Chapter 5, Feller's recent treatment of diffusion processes was discussed by infusing with the methods developed by E. B. Dynkin. There are many results from olden days concerning ergodic properties of Markov processes, but we omitted discussions of this topic, because of the limitation of the pages available.

Classical results on this topic are summarized in

- M. Fréchet: Recherches théoretiques modernes sur le calcul des probabilités, second livre, methode des fonctions arbitraires, théorie des événement en chain dans le cas d'un nombre fini d'étas possibles (Paris, 1938).

This treats, as the title of the book indicates, the case of finite state Markov processes. A pioneering work on Markov processes with countable infinity of states is the following article by Lévy:

- P. Lévy: Systèms markoviens et stationaires. Cas dénombrable. Ann. Sci. École Norm. Sup. 68 (1951).

Chung has published many results obtained by detailed studies trying to make the results in this article by Lévy more rigorous. The aforementioned book by G. Maruyama provides a good account in Japanese in this area.

There are many investigations for temporally non-homogeneous Markov processes as well. We refer the readers to the aforementioned books by Feller, Itô, and Doob on this topic. We recommend the readers to go through the following work of Kolmogorov, which started the studies in this direction:

- A. Kolmogoroff: Analytische Methoden in der Wahrscheinlichkeitsrechnung, Math. Ann. 104.

Titles in This Series

231 **Kiyosi Itô,** Essentials of stochastic processes, 2006

230 **Akira Kono and Dai Tamaki,** Generalized cohomology, 2006

229 **Yu. N. Lin′kov,** Lectures in mathematical statistics, 2005

228 **D. Zhelobenko,** Principal structures and methods of representation theory, 2006

227 **Takahiro Kawai and Yoshitsugu Takei,** Algebraic analysis of singular perturbation theory, 2005

226 **V. M. Manuilov and E. V. Troitsky,** Hilbert C^*-modules, 2005

225 **S. M. Natanzon,** Moduli of Riemann surfaces, real algebraic curves, and their superanaloges, 2004

224 **Ichiro Shigekawa,** Stochastic analysis, 2004

223 **Masatoshi Noumi,** Painlevé equations through symmetry, 2004

222 **G. G. Magaril-Il'yaev and V. M. Tikhomirov,** Convex analysis: Theory and applications, 2003

221 **Katsuei Kenmotsu,** Surfaces with constant mean curvature, 2003

220 **I. M. Gelfand, S. G. Gindikin, and M. I. Graev,** Selected topics in integral geometry, 2003

219 **S. V. Kerov,** Asymptotic representation theory of the symmetric group and its applications to analysis, 2003

218 **Kenji Ueno,** Algebraic geometry 3: Further study of schemes, 2003

217 **Masaki Kashiwara,** D-modules and microlocal calculus, 2003

216 **G. V. Badalyan,** Quasipower series and quasianalytic classes of functions, 2002

215 **Tatsuo Kimura,** Introduction to prehomogeneous vector spaces, 2003

214 **L. Š. Grinblat,** Algebras of sets and combinatorics, 2002

213 **V. N. Sachkov and V. E. Tarakanov,** Combinatorics of nonnegative matrices, 2002

212 **A. V. Mel′nikov, S. N. Volkov, and M. L. Nechaev,** Mathematics of financial obligations, 2002

211 **Takeo Ohsawa,** Analysis of several complex variables, 2002

210 **Toshitake Kohno,** Conformal field theory and topology, 2002

209 **Yasumasa Nishiura,** Far-from-equilibrium dynamics, 2002

208 **Yukio Matsumoto,** An introduction to Morse theory, 2002

207 **Ken'ichi Ohshika,** Discrete groups, 2002

206 **Yuji Shimizu and Kenji Ueno,** Advances in moduli theory, 2002

205 **Seiki Nishikawa,** Variational problems in geometry, 2001

204 **A. M. Vinogradov,** Cohomological analysis of partial differential equations and Secondary Calculus, 2001

203 **Te Sun Han and Kingo Kobayashi,** Mathematics of information and coding, 2002

202 **V. P. Maslov and G. A. Omel′yanov,** Geometric asymptotics for nonlinear PDE. I, 2001

201 **Shigeyuki Morita,** Geometry of differential forms, 2001

200 **V. V. Prasolov and V. M. Tikhomirov,** Geometry, 2001

199 **Shigeyuki Morita,** Geometry of characteristic classes, 2001

198 **V. A. Smirnov,** Simplicial and operad methods in algebraic topology, 2001

197 **Kenji Ueno,** Algebraic geometry 2: Sheaves and cohomology, 2001

196 **Yu. N. Lin′kov,** Asymptotic statistical methods for stochastic processes, 2001

195 **Minoru Wakimoto,** Infinite-dimensional Lie algebras, 2001

194 **Valery B. Nevzorov,** Records: Mathematical theory, 2001

193 **Toshio Nishino,** Function theory in several complex variables, 2001

192 **Yu. P. Solovyov and E. V. Troitsky,** C^*-algebras and elliptic operators in differential topology, 2001

191 **Shun-ichi Amari and Hiroshi Nagaoka,** Methods of information geometry, 2000

190 **Alexander N. Starkov,** Dynamical systems on homogeneous spaces, 2000

189 **Mitsuru Ikawa,** Hyperbolic partial differential equations and wave phenomena, 2000

188 **V. V. Buldygin and Yu. V. Kozachenko,** Metric characterization of random variables and random processes, 2000

187 **A. V. Fursikov,** Optimal control of distributed systems. Theory and applications, 2000

186 **Kazuya Kato, Nobushige Kurokawa, and Takeshi Saito,** Number theory 1: Fermat's dream, 2000

185 **Kenji Ueno,** Algebraic Geometry 1: From algebraic varieties to schemes, 1999

184 **A. V. Mel′nikov,** Financial markets, 1999

183 **Hajime Sato,** Algebraic topology: an intuitive approach, 1999

182 **I. S. Krasil′shchik and A. M. Vinogradov, Editors,** Symmetries and conservation laws for differential equations of mathematical physics, 1999

181 **Ya. G. Berkovich and E. M. Zhmud′,** Characters of finite groups. Part 2, 1999

180 **A. A. Milyutin and N. P. Osmolovskii,** Calculus of variations and optimal control, 1998

179 **V. E. Voskresenskiĭ,** Algebraic groups and their birational invariants, 1998

178 **Mitsuo Morimoto,** Analytic functionals on the sphere, 1998

177 **Satoru Igari,** Real analysis—with an introduction to wavelet theory, 1998

176 **L. M. Lerman and Ya. L. Umanskiy,** Four-dimensional integrable Hamiltonian systems with simple singular points (topological aspects), 1998

175 **S. K. Godunov,** Modern aspects of linear algebra, 1998

174 **Ya-Zhe Chen and Lan-Cheng Wu,** Second order elliptic equations and elliptic systems, 1998

173 **Yu. A. Davydov, M. A. Lifshits, and N. V. Smorodina,** Local properties of distributions of stochastic functionals, 1998

172 **Ya. G. Berkovich and E. M. Zhmud′,** Characters of finite groups. Part 1, 1998

171 **E. M. Landis,** Second order equations of elliptic and parabolic type, 1998

170 **Viktor Prasolov and Yuri Solovyev,** Elliptic functions and elliptic integrals, 1997

169 **S. K. Godunov,** Ordinary differential equations with constant coefficient, 1997

168 **Junjiro Noguchi,** Introduction to complex analysis, 1998

167 **Masaya Yamaguti, Masayoshi Hata, and Jun Kigami,** Mathematics of fractals, 1997

166 **Kenji Ueno,** An introduction to algebraic geometry, 1997

165 **V. V. Ishkhanov, B. B. Lur′e, and D. K. Faddeev,** The embedding problem in Galois theory, 1997

164 **E. I. Gordon,** Nonstandard methods in commutative harmonic analysis, 1997

163 **A. Ya. Dorogovtsev, D. S. Silvestrov, A. V. Skorokhod, and M. I. Yadrenko,** Probability theory: Collection of problems, 1997

For a complete list of titles in this series, visit the AMS Bookstore at **www.ams.org/bookstore/**.